"工学结合、校企合作"课程改革系列教材
全国职业院校技能大赛计算机类项目辅导用书

网络组建与管理实训教程

丛书主编　吴访升
主　　编　韩红章
副 主 编　朱林立　臧海娟
参　　编　狄涵妤　王　尧
丛书主审　董群朴

机械工业出版社

本书是以实际工程项目为主体，以实践为主线编写的模块化教材，采用分步解决各个子项目问题的方法来介绍网络组建与管理的主要内容。全书共7章，包括网络工程项目描述及需求分析、交换技术、路由技术、网络安全技术、远程接入技术、网络服务应用和网络工程项目规划与搭建。

本书将网络组建与管理通过一个学校的实际网络工程项目分解为若干个任务进行介绍，将网络组建与管理的理念和技术逐次展开，让读者在具体案例的操练中逐步构建和提升相关理论知识。

本书适用于职业院校计算机及相关专业的学生、省级和国家级计算机技能比赛选手和教练员以及网络组建与管理的爱好者。

本书配套电子课件，选择本书作为教材的教师可到机械工业出版社教材服务网www.cmpedu.com注册并下载，或联系编辑（010-88379194）索取。

图书在版编目（CIP）数据

网络组建与管理实训教程/韩红章主编．—北京：机械工业出版社，2015.1（2022.8重印）
"工学结合、校企合作"课程改革系列教材
全国职业院校技能大赛计算机类项目辅导用书
ISBN 978-7-111-49742-4

Ⅰ．①网… Ⅱ．①韩… Ⅲ．①计算机网络—高等职业教育—教材 Ⅳ．①TP393

中国版本图书馆CIP数据核字（2015）第057727号

机械工业出版社（北京市百万庄大街22号 邮政编码100037）
策划编辑：梁 伟　　责任编辑：蔡 岩
封面设计：鞠 杨　　责任校对：朱继文
责任印制：常天培
固安县铭成印刷有限公司印刷
2022年8月第1版第6次印刷
184mm×260mm・14印张・329千字
标准书号：ISBN 978-7-111-49742-4
定价：45.00元

电话服务　　　　　　　　　　网络服务
客服电话：010-88361066　　机 工 官 网：www.cmpbook.com
　　　　　010-88379833　　机 工 官 博：weibo.com/cmp1952
　　　　　010-68326294　　金 书 网：www.golden-book.com
封底无防伪标均为盗版　　　　机工教育服务网：www.cmpedu.com

前　言

本书打破了传统的以知识传授为主线的知识架构，采用案例和项目设计相结合的编写方法，将网络组建与管理通过一个学校的实际网络工程项目分解为若干个任务进行介绍，将网络组建与管理的理念和技术逐次展开，让学生在具体案例的操练中逐步构建和掌握相关理论知识。通过项目的活动过程可以培养学生的分析问题能力、团队精神和沟通能力。

本书围绕网络组建与管理的职业要求，以针对中等职业学校计算机类专业学生的学习特点、培养目标和要求为依据，紧扣全国职业院校技能大赛最新精神和动态，教材内容力求真正体现以教师为主导、学生为主体的教学理念，来激发学生的学习兴趣，充分发掘学生的学习潜能。本书在项目选择上力求科学、全面、实用，有助于学生加深对知识点的理解。本书表述形式新颖、生动，多采用图文结合的形式，力求直观明了。

本书共分为7章，主要内容如下：

第0章为网络工程项目描述及需求分析，主要描述项目具体需求。

第1章为交换技术，包括VLAN技术、链路聚合技术和（STP、RSTP、MSTP）生成树技术配置与管理等。

第2章为路由技术，包括静态路由、RIP路由、OSPF路由、路由选择控制及路由重分发和策略路由配置与管理等。

第3章为网络安全技术，包括交换机端口安全、DHCP监听、ARP检查、ACL应用和网络地址转换配置与管理等。

第4章为远程接入技术，包括点对点协议PPP和帧中继技术配置与管理等。

第5章为网络服务应用，包括Telnet服务、DHCP服务、VRRP技术、VPN技术、QoS技术、组播技术和IPv6技术配置与管理等。

第6章为网络工程项目规划与搭建，主要是根据单位的具体需求，对网络工程项目进行合理规划以及分项目逐步搭建。

本系列丛书由吴访升任丛书主编，董群朴任丛书主审。本书由韩红章任主编，朱林立、臧海娟任副主编，参与编写的还有浃涵妤和王尧。

由于编者水平所限，教材内容难免有疏漏和不当之处，恳请各位专家、学校师生及广大读者批评指正。

编　者

目 录

前言
第0章 网络工程项目描述及需求分析 1
第1章 交换技术 5
1.1　VLAN技术 6
1.2　端口聚合技术 12
1.3　生成树技术 17
第2章 路由技术 27
2.1　静态路由 28
2.2　RIP V2路由 34
2.3　OSPF路由 41
2.4　路由重分发与路由控制 53
2.5　策略路由 60
第3章 网络安全技术 70
3.1　端口安全 71
3.2　DHCP监听 73
3.3　ARP检查 76
3.4　ACL应用 80
3.5　网络地址转换技术 88
第4章 远程接入技术 96
4.1　点对点协议PPP 97
4.2　帧中继技术 103
第5章 网络服务应用 111
5.1　Telnet服务 112
5.2　DHCP服务 116
5.3　VRRP技术 123
5.4　VPN技术 131
5.5　QoS技术 153
5.6　组播技术 162
5.7　IPv6技术 168
第6章 网络工程项目规划与搭建 173
参考文献 220

第0章　网络工程项目描述及需求分析

项目描述

某中等职业学校现有师生员工5000余人，分布在两个不同的校区，为了实现快捷的信息交流和资源共享，需要构建一个跨越两地的校园网络。主校区主要有教学中心、实训中心、行政办公中心、信息中心（服务器群）四大区域，分校区主要有教学中心、实训中心、行政办公中心、无线区域四大区域。单位内部网络规划拥有自己的域名服务器、邮件服务器、Web服务器、FTP服务器和VPN服务器。

主校区采用双核心的网络架构和双出口的网络接入模式，主校区对外采用100Mbit/s电信链路和1000Mbit/s教育网链路。校内双核心交换机之间采用链路聚合技术，增加网络的带宽并可以实现流量负载均衡。

分校区采用路由器接入互联网和城域网的模式，接入互联网为100Mbit/s电信链路。分校区的行政办公中心用户采用有线和无线两种接入方式，更便于访问网络资源。

为了保障主校区与分校区业务数据流传输的高可用性，采用城域网专用链路为主链路，采用基于IPSec VPN技术的互联网链路为备份链路，实现业务流量的高可用性。

为了保障主校区网络的稳定性和拓扑快速收敛，在IP选路中采用的是动态路由协议，因为网络规模较大，所以采用的动态路由协议是开放式最短路径优先（OSPF）。为保障路由协议的更新，需要进行基于接口的MD5验证。

为了实现资源的共享及信息的发布，主校区信息中心搭建了应用服务器群。为了实现快捷的信息传递和学校业务的需求，允许SOHO办公和出差的教职员工能够方便、快捷、安全地访问学校内网服务器群。

要求对局域网进行合理网络规划，要求各大区域能使区域内所有设备均能在通过学校的网关后访问Internet资源，在局域网内部能够实现互联，实现全网的互通。在网络建设中，保证整个网络系统的高性能、高可靠性、高稳定性、高灵活性和高综合性。

项目分析

网络工程所需要规划的项目非常多，也非常广，涉及各具体项目之间的关联与综合考虑。根据该单位的要求，我们利用网络层次化结构模型对网络进行合理的设计、规划与管理来解决上述问题。可将该项目分解为4个部分：①根据单位的实际需求，利用交换技术和路由技术,对基础网络进行规划和搭建,实现局域网的互联互通;②根据单位的实际需求,建立一套有效的网络安全机制,保护内部网络的安全;③利用远程接入技术,实现局域网与广域网的互联;④在网络互联互通的基础上,进行网络服务应用的搭建,提高网络系统的高可靠性、高稳定性、高灵活性和高综合性。

1. 基础网络规划和搭建

1)该校内部有多个部门,采用VLAN技术,将多个部门的主机划分到不同的VLAN中,既可以实现统一管理,又可以保障网络的安全。

2)为了保障网络流量畅通,在核心交换机之间的链路中采用链路聚合技术,实现链路带宽的增加和负载均衡;在接入层二层交换机上行至三层交换的链路中采用链路冗余技术,防止链路中断,形成单点故障。

3)在三层路由规划时,需要利用静态或动态路由实现内网互通;如果网络中存在不同的路由协议,需要使用重发布技术,实现全网互通。

2. 网络安全搭建

1)在保障接入层安全方面主要是使用端口安全、DHCP监听、ARP检测等技术来实现。

2)在保护网络资源方面主要是使用访问控制列表和网络地址转换等技术来实现。

3. 远程接入技术

远程接入技术主要是通过PPP协议和帧中继技术来实现的。

4. 网络服务应用

Telnet服务、DHCP服务、VRRP技术、VPN技术、QoS技术、组播技术和IPv6技术等网络服务的应用主要是优化网络资源,提高网络服务性能,为用户提供最优质的网络服务。

全局规划

1. 网络拓扑结构规划(见图0-1)

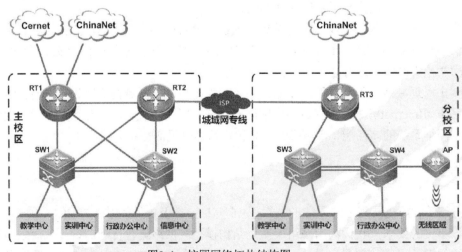

图0-1　校园网络拓扑结构图

第0章 网络工程项目描述及需求分析

说明：在网络拓扑结构规划中，没有把核心层设备和汇聚层设备按照实际情况分开来，而是把核心层和汇聚层合为一层，主要是为了简化拓扑结构以及实验的方便。拓扑结构图中教学中心、实训中心等区域使用接入层设备连接。

2. 部门名称规划

主校区教学中心为VLAN 10、实训中心为VLAN 20、行政办公中心为VLAN 30、信息中心为VLAN 100；分校区教学中心为VLAN 40、实训中心为VLAN 50、行政办公中心为VLAN 60、无线办公区域为VLAN 200。

3. IP地址规划（见表0-1）

表0-1 IP地址规划表

网络设备名称	ID	IP地址	备注
RT1	F0/0	10.1.1.1/30	
	F0/1	10.1.1.5/30	
	S2/0	10.1.1.9/30	
	S3/0	210.21.1.2/24	
	S4/0	68.1.1.2/28	
RT2	F0/0	10.1.1.13/30	
	F0/1	10.1.1.17/30	
	S2/0	10.1.1.10/30	
	S3/0	10.1.1.25/30	
RT3	F0/0	10.1.1.29/30	
	F0/1	10.1.1.33/30	
	S2/0	72.1.1.2/30	
	S3/0	10.1.1.26/30	
SW1	F0/1	10.1.1.2/30	
	F0/2	10.1.1.14/30	
	F0/23-24	10.1.1.21/30	
	VLAN 10	192.168.10.1/24	F0/6-10
	VLAN 20	192.168.20.1/24	F0/11-15
SW2	F0/1	10.1.1.6/30	
	F0/2	10.1.1.18/30	
	F0/23-24	10.1.1.22/30	
	VLAN 30	192.168.30.1/24	F0/6-10
	VLAN 100	192.168.100.1/24	F0/11-15
SW3	F0/1	10.1.1.30/30	
	F0/23-24	10.1.1.41/30	
	VLAN 40	192.168.40.1/24	F0/6-10
	VLAN 50	192.168.50.1/24	F0/11-15

（续）

网络设备名称	ID	IP地址	备注
SW4	F0/1	10.1.1.34/30	
	F0/22	10.1.1.37/30	
	F0/23-24	10.1.1.42/30	
AP	VLAN 60	192.168.60.1/24	F0/6-15
	VLAN 200	192.168.200.1/24	无线区域
	G0/1	10.1.1.38/30	

4. 本书中使用的命令语法规范说明

- ◇ 竖线"|"：表示分隔符，用于分开可选择的选项。
- ◇ 方括号"[]"：表示可选项。
- ◇ 大括号"{}"：表示必选项。
- ◇ 粗体字：表示按照显示输入的文字、输入的命令和关键字。
- ◇ 斜体字：表示需要用户输入的具体值。

第1章 交换技术

随着互联网技术的迅猛发展，交换技术也从传统的电路交换、分组交换发展到现在的以IP为核心的宽带分组交换，再到光交换。交换技术已经占据了目前网络的主导地位。

从分层次的模块化网络结构来看，在当前校园网络中部署最多的网络组件是提供信息节点接入的二层交换机以及提供汇聚和核心转发的三层交换机。并且，三层交换机的出现解决了低端路由器的端口密度小、性能差等缺点，使路由选择和高性能的交换结合到一起，为校园网络提供了高性能的、灵活的、可靠的解决方案。

本章介绍的交换技术主要有VLAN技术、端口聚合技术和生成树技术。

1. 虚拟局域网（VLAN）

VLAN是一种可以把局域网内的交换设备逻辑地而不是物理地划分成一个个网段的技术，也就是从物理网络上划分出来的逻辑网络。由于VLAN是基于逻辑连接而不是物理连接，所以它可以提供灵活的用户管理、带宽分配以及资源优化等服务。

VLAN的划分可以依据网络用户的组织结构进行，形成一个个虚拟的工作组。这样，网络中的工作组就可以突破共享网络中地理位置的限制，而根据管理功能来划分。这种基于工作流的分组模式，很好地提高了网络的管理功能。

2. 端口聚合技术

端口聚合技术是指可以把多个物理端口捆绑在一起形成一个简单的逻辑端口，这个逻辑端口被称为端口聚合。

端口聚合技术可以为高带宽网络连接实现负载共享、负载平衡以及提供更好的弹性，而且由于逻辑上是单个端口，所以也不存在环路问题。这项链路聚合标准在点到点链路上提供了固有的、自动的冗余性。换句话说，如果链路中所使用的多个端口中的一个端口出现故障，网络传输流可以动态地改向链路中余下的正常状态的端口进行传输。

3. 生成树技术

生成树技术的主要实现原理就是当网络中存在备份链路时，只允许主链路激活。如果主链路因故障而被断开，则备用链路才会被打开。

当交换机间存在多条链路时，交换机的生成树算法只启动最主要的一条链路，而将其他链路都阻塞掉，并变为备用链路。当主链路出现问题时，生成树协议将自动起用备用链路接替主链路工作，不需要任何人干预。

1.1 VLAN技术

问题描述

学校内部有多个部门，根据部门业务的不同进行VLAN区划，既可以实现统一管理，又可以保障网络的安全，为了便于网络管理，每个VLAN按照部门名称的汉语拼音进行命名。

主校区主要有教学中心、实训中心、行政办公中心、信息中心（服务器群）四大区域。创建VLAN 10、VLAN 20、VLAN 30和VLAN 100，将教学中心的用户主机划分到VLAN 10，将实训中心的用户主机划分到VLAN 20，将行政办公中心的用户主机划分到VLAN 30，将信息中心的服务器群划分到VLAN 100；其中VLAN 10对应的IP地址为192.168.10.1/24，VLAN 20对应的IP地址为192.168.20.1/24，VLAN 30对应的IP地址为192.168.30.1/24，VLAN 100对应的IP地址为192.168.100.1/24。

分校区主要有教学中心、实训中心、行政办公中心、无线区域四大区域。创建VLAN 40、VLAN 50、VLAN 60和VLAN 200，将教学中心的用户主机划分到VLAN 40，将实训中心的用户主机划分到VLAN 50，将行政办公中心的用户主机划分到VLAN 60，将无线办公的用户主机划分到VLAN 200；其中VLAN 40对应的IP地址为192.168.40.1/24，VLAN 50对应的IP地址为192.168.50.1/24，VLAN 60对应的IP地址为192.168.60.1/24，VLAN 200对应的IP地址为192.168.200.1/24。

问题分析

VLAN的划分主要是VLAN的创建、端口的隔离、配置Trunk口、开启三层交换机端口路由与VLAN间路由。

1. VLAN创建

步骤1 创建VLAN。vlan-id是VLAN的编号。

switch(config)#**vlan** *vlan-id*

步骤2 命名VLAN（可选）。vlan-name是VLAN的名字，主要方便网络管理员的管理。

switch(config-vlan)#**name** *vlan-name*

2. 将交换机端口划分到VLAN中

步骤1 进入需要配置的端口。interface-id是端口的编号，如果有多个接口要加入到同一个VLAN，用interface range {port-range}来批量设置接口。

switch(config)#**interface** *interface-id*

步骤2 定义该接口的VLAN成员类型是二层Access接口。

switch(config-if)#switchport mode access

步骤3 将端口添加到特定的VLAN中。

switch(config-if)#**switchport access vlan** *vlan-id*

3. 将级联端口设置为Trunk

步骤1 进入需要配置的端口。

switch(config)#**interface** *interface-id*

第1章 交换技术

步骤2 将端口的模式设置为Trunk。

switch(config-if)#**switchport mode trunk**

4. 配置SVI

步骤1 进入VLAN的SVI接口配置模式。

switch(config)#**interface vlan** *vlan-id*

步骤2 给SVI接口配置IP地址。这些IP地址将作为各个VLAN内主机的网关，并且这些虚拟接口所在的网段也会作为直接路由出现在三层交换机的路由表中。ip-address是需要配置的IP地址，mask为子网掩码。

switch(config-if)#**ip address** *ip-address mask*

5. 配置单臂路由

步骤1 创建以太网子接口。interface-id.sub-port是子接口的编号。

router(config)#**interface** *interface-id.sub-port*

步骤2 为子接口封装802.1q协议，并指定接口所属的VLAN。

router(config-subif)#**encapsulation dot1q** *vlan-id*

步骤3 为子接口配置IP地址。

switch(config-subif)#**ip address** *ip-address mask*

步骤4 启用子接口。

switch(config-subif)#**no shutdown**

任 务 单

1	使用SVI实现VLAN间通信
2	跨交换机实现VLAN间通信
3	使用单臂路由实现VLAN间通信

解决步骤

根据任务单的安排完成任务。

任务1：使用SVI实现VLAN间通信

任务实施

1. 任务描述及网络拓扑设计

在SW1三层交换机上为VLAN 10和VLAN 20配置SVI虚拟接口，利用三层交换机的路由功能实现VLAN间路由。绘制拓扑结构图，如图1-1所示。

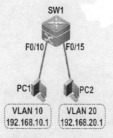

图1-1 使用SVI实现VLAN间通信

2. 网络设备配置

（1）在SW1交换机上创建VLAN、端口隔离、配置SVI

SW1>enable	#进入特权模式
SW1#configure terminal	#进入全局模式
SW1(config)#vlan 10	#创建VLAN 10
SW1(config-vlan)#exit	#返回全局模式
SW1(config)#vlan 20	#创建VLAN 20
SW1(config-vlan)#exit	#返回全局模式
SW1(config)#interface range fastEthernet 0/6-10	#进入接口模式
SW1(config-if-range)#switchport access vlan 10	#将接口划入到VLAN 10
SW1(config-if-range)#exit	#返回全局模式
SW1(config)#interface range fastEthernet 0/11-15	#进入接口模式
SW1(config-if-range)#switchport access vlan 20	#将接口划入到VLAN 20
SW1(config-if-range)#exit	#返回全局模式
SW1(config)#interface vlan 10	#进入VLAN接口模式
SW1(config-VLAN 10)#ip address 192.168.10.1 255.255.255.0	#配置SVI接口地址
SW1(config-VLAN 10)#exit	
SW1(config)#interface vlan 20	#进入VLAN接口模式
SW1(config-VLAN 20)#ip address 192.168.20.1 255.255.255.0	#配置SVI接口地址
SW1(config-VLAN 20)#exit	

（2）测试网络连通性

按图1-1连接拓扑，给PC1主机配置相应的IP地址为192.168.10.2/24，网关为192.168.10.1；给PC2主机配置相应的IP地址为192.168.20.2/24，网关为192.168.20.1。从VLAN 10中的PC1 ping VLAN 20中的PC2，结果如下所示。

```
C:\Documents and Settings\Administrator>ping 192.168.20.2

Pinging 192.168.20.2 with 32 bytes of data:

Reply from 192.168.20.2: bytes=32 time<1ms TTL=63
Reply from 192.168.20.2: bytes=32 time<1ms TTL=63
Reply from 192.168.20.2: bytes=32 time<1ms TTL=63
Reply from 192.168.20.2: bytes=32 time<1ms TTL=63

Ping statistics for 192.168.20.2:
    Packets: Sent = 4, Received = 4, Lost = 0 (0% loss),
Approximate round trip times in milli-seconds:
    Minimum = 0ms, Maximum = 0ms, Average = 0ms
```

任务2：跨交换机实现VLAN间通信

任务实施

1. 任务描述及网络拓扑设计

在SWA与SWB二层交换机上分别创建VLAN 40和VLAN 50，配置Trunk实现不同VLAN主机接入；在SW3三层交换机上创建VLAN 40和VLAN 50，并且配置SVI虚拟接口，利用三层交换机的路由功能实现不同的VLAN间路由。绘制拓扑结构图，如图1-2所示。

第1章 交换技术

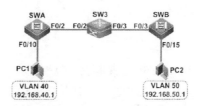

图1-2 跨交换机实现VLAN间通信

2. 网络设备配置

（1）在SW3三层交换机上创建VLAN、配置SVI，配置端口trunk模式

SW3>enable	#从用户模式进入特权模式
SW3#configure terminal	#从特权模式进入全局模式
SW3(config)#vlan 40	#创建VLAN 40
SW3(config-vlan)#exit	#返回全局模式
SW3(config)#vlan 50	#创建VLAN 50
SW3(config-vlan)#exit	#返回全局模式
SW3(config)#interface vlan 40	#进入VLAN接口模式
SW3(config-VLAN 40)#ip address 192.168.40.1 255.255.255.0	#配置SVI接口地址
SW3(config-VLAN 40)#exit	
SW3(config)#interface vlan 50	
SW3(config-VLAN 50)#ip address 192.168.50.1 255.255.255.0	#配置SVI接口地址
SW3(config-VLAN 50)#exit	
SW3(config)#interface fastEthernet 0/2	#进入接口模式
SW3(config-FastEthernet 0/2)#switchport mode trunk	#设置端口模式为trunk
SW3(config-FastEthernet 0/2)#exit	
SW3(config)#interface fastEthernet 0/3	#进入接口模式
SW3(config-FastEthernet 0/3)#switchport mode trunk	#设置端口模式为trunk

（2）在SWA二层交换机上创建VLAN、端口隔离

```
SWA>enable
SWA#configure terminal
SWA(config)#vlan 40
SWA(config-vlan)#exit
SWA(config)#interface fastEthernet 0/10
SWA(config-FastEthernet 0/10)#switchport access vlan 40
SWA(config-FastEthernet 0/10)#exit
SWA(config)#interface fastEthernet 0/2
SWA(config-FastEthernet 0/2)#switchport mode trunk
SWA(config-FastEthernet 0/2)#exit
SWA(config)#exit
```

（3）在SWB二层交换机上创建VLAN、端口隔离

```
SWB>enable
SWB#configure terminal
SWB(config)#vlan 50
SWB(config-vlan)#exit
SWB(config)#interface fastEthernet 0/15
SWB(config-FastEthernet 0/15)#switchport access vlan 50
SWB(config-FastEthernet 0/15)#exit
SWB(config)#interface fastEthernet 0/3
SWB(config-FastEthernet 0/3)#switchport mode trunk
```

SWB(config-FastEthernet 0/3)#exit
SWB(config)#exit

（4）测试网络连通性

按图1-2连接拓扑，给PC1主机配置相应的IP地址为192.168.40.2/24，网关为192.168.40.1；给PC2主机配置相应的IP地址为192.168.50.2/24，网关为192.168.50.1，从VLAN 40中的PC1 ping VLAN 50中的PC2，结果如下所示。

C:\Documents and Settings\Administrator>ping 192.168.50.2

Pinging 192.168.50.2 with 32 bytes of data:

Reply from 192.168.50.2: bytes=32 time<1ms TTL=63
Reply from 192.168.50.2: bytes=32 time<1ms TTL=63
Reply from 192.168.50.2: bytes=32 time<1ms TTL=63
Reply from 192.168.50.2: bytes=32 time<1ms TTL=63

Ping statistics for 192.168.50.2:
    Packets: Sent = 4, Received = 4, Lost = 0 (0% loss),
Approximate round trip times in milli-seconds:
    Minimum = 0ms, Maximum = 0ms, Average = 0ms

任务3：使用单臂路由实现VLAN间通信

任务实施

1. 任务描述及网络拓扑设计

在RTA上对物理接口划分子接口并封装802.1Q协议，使得每一个子接口分别充当VLAN10和VLAN20网段中主机的网关，利用路由器的直连路由功能实现不同VLAN间的通信。绘制拓扑结构图，如图1-3所示。

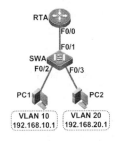

图1-3　使用单臂路由实现VLAN间通信

2. 网络设备配置

（1）在SWA交换机上创建VLAN、端口隔离

SWA(config)#vlan 10
SWA(config-vlan)#exit
SWA(config)#vlan 20
SWA(config-vlan)#exit
SWA(config)#interface fastethernet 0/1
SWA(config-FastEthernet 0/1)#switchport mode trunk
SWA(config-FastEthernet 0/1)#exit

第1章 交换技术

```
SWA(config)#interface fastEthernet 0/2
SWA(config-FastEthernet 0/2)#switchport access vlan 10
SWA(config-FastEthernet 0/2)#exit
SWA(config)#interface fastEthernet 0/3
SWA(config-FastEthernet 0/3)#switchport access vlan 20
SWA(config-FastEthernet 0/3)#exit
```

（2）在RTA路由器上配置单臂路由

```
RTA(config)#interface fastEthernet 0/0
RTA(config-if-FastEthernet 0/0)#no shutdown
RTA(config)#interface fastEthernet 0/0.1              #进入子接口模式
RTA(config-subif)#encapsulation dot1Q 10              #封装VLAN标签
RTA(config-subif)#ip address 192.168.10.1 255.255.255.0   #配置子接口IP地址
RTA(config-subif)#no shutdown                         #启用子接口
RTA(config-subif)#exit
RTA(config)#interface fastEthernet 0/0.2              #进入子接口模式
RTA(config-subif)#encapsulation dot1Q 20              #封装802.1q协议
RTA(config-subif)#ip address 192.168.20.1 255.255.255.0   #配置子接口IP地址
RTA(config-subif)#no shutdown                         #启用子接口
```

（3）测试网络连通性

按图1-3连接拓扑，给PC1主机配置相应的IP地址为192.168.10.2/24，网关为192.168.10.1；给PC2主机配置相应的IP地址为192.168.20.2/24，网关为192.168.20.1，从VLAN 10中的PC1 ping VLAN 20中的PC2，结果如下所示。

```
C:\Documents and Settings\Administrator>ping 192.168.20.2

Pinging 192.168.20.2 with 32 bytes of data:

Reply from 192.168.20.2: bytes=32 time<1ms TTL=63
Reply from 192.168.20.2: bytes=32 time<1ms TTL=63
Reply from 192.168.20.2: bytes=32 time<1ms TTL=63
Reply from 192.168.20.2: bytes=32 time<1ms TTL=63
Ping statistics for 192.168.20.2:
    Packets: Sent = 4, Received = 4, Lost = 0 (0% loss),
Approximate round trip times in milli-seconds:
    Minimum = 0ms, Maximum = 0ms, Average = 0ms
```

小结

通过VLAN技术的学习，主要掌握VLAN的创建、端口的隔离、Trunk的划分以及开启三层交换机的路由功能、VLAN间路由等配置与管理工作。由于VLAN隔离了广播域，所以要实现VLAN之间的通信需要三层设备的支持，可通过路由器直连路由或者通过三层交换机以SVI接口的方式来实现。

VLAN 1属于系统默认的VLAN，不可以被删除。删除某个VLAN时，应先将属于该VLAN的端口加入到其他VLAN中，再对其进行删除。

交换机端口分为Access口与Trunk口两种类型，Access口类型表示一个端口只属于一个VLAN，Trunk口类型表示一个端口属于该交换设备上创建的所有VLAN，主要用于实现跨交换机的相同VLAN内主机之间可以直接访问。

1.2 端口聚合技术

问题描述

随着该校各个部门对网络的依赖越来越强，为了保证网络的高可用性，有时会希望在网络中提供设备、模块和链路的冗余。

主校区采用双核心网络架构，为了保障网络链路的带宽，将两台三层交换机相连接的链路配置为链路聚合，并使用链路负载均衡技术，使链路的流量基于源IP地址负载均衡。为了保障连接服务器群二层链路的冗余和负载均衡，需要配置二层端口聚合技术，并使用基于目的MAC地址的流量平衡。

问题分析

端口聚合技术主要涉及二层端口聚合、三层端口聚合以及配置流量平衡。

1. 配置二层Aggregate Port

步骤1 创建聚合端口AP，其中aggregate-port-number为AP号。

switch(config)#**interface aggregateport** *aggregate-port-number*

步骤2 选择端口，进入接口配置模式，指定要加入AP的物理端口范围。

switch(config)# **interface range** *port-range*

步骤3 将该端口加入一个AP。

switch(config-if-range)#**port-group** *aggregate-port-number*

2. 配置三层Aggregate Port

步骤1 创建聚合端口AP。

switch(config)#**interface aggregateport** *aggregate-port-number*

步骤2 将该端口设置为三层模式。

switch(config-if)#**no switchport**

步骤3 给AP端口设置IP地址和子网掩码。

switch(config-if)#**ip address** *ip-address mask*

3. 配置流量平衡

设置AP的流量平衡，选择使用的算法。其中dst-mac表示根据输入报文的目的MAC地址进行流量分配，src-mac表示根据输入报文的源MAC地址进行流量分配，src-dst-mac表示根据源MAC与目的MAC进行流量分配，dst-ip表示根据输入报文的目的IP地址进行流量分配，src-ip表示根据输入报文的源IP地址进行流量分配，ip表示根据源IP和目的IP进行流量分配。

switch(config)#**aggregateport load-balance** {dst-mac/src-mac/src-dst-mac/dst-ip/src-ip/ip}

任 务 单

1	配置二层端口聚合
2	配置三层端口聚合

第1章 交换技术

根据任务单的安排完成任务。

▶ **任务1：配置二层端口聚合**

任务实施

1. 任务描述及网络拓扑设计

在SWA和SWB两台二层交换机之间的冗余链路上实现端口聚合，且在聚合端口上实现基于目的MAC地址的流量负载均衡。绘制拓扑结构图，如图1-4所示。

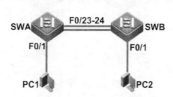

图1-4 二层端口聚合

2. 网络设备配置

（1）在SWA交换机上创建端口聚合，并设置为Trunk模式

```
SWA(config)#interface aggregateport 1                    #创建端口聚合1
SWA(config-AggregatePort 1)#exit
SWA(config)#interface range fastEthernet 0/23-24         #进入一组接口
SWA(config-if-range)#port-group 1                        #将接口配置成AP的成员端口
SWA(config-if-range)#exit
SWA(config)#interface aggregateport 1                    #进入聚合接口模式
SWA(config-AggregatePort 1)#switchport mode trunk        #将端口聚合配置为干道模式
```

（2）在SWB交换机上创建端口聚合，并设置为Trunk模式

```
SWB(config)#interface aggregateport 1                    #创建端口聚合1
SWB(config-AggregatePort 1)#exit
SWB(config)#interface range fastEthernet 0/23-24         #进入一组接口
SWB(config-if-range)#port-group 1                        #将接口配置成AP的成员端口
SWB(config-if-range)#exit
SWB(config)#interface aggregateport 1                    #进入聚合接口模式
SWB(config-AggregatePort 1)#switchport mode trunk        #将端口聚合配置为干道模式
```

（3）在SWA交换机上的端口聚合上配置流量平衡

```
SWA(config)#aggregateport load-balance ?                 #查看流量负载均衡的方式
  dst-ip       Destination IP address
  dst-mac      Destination MAC address
  src-dst-ip   Source and destination IP address
  src-dst-mac  Source and destination MAC address
  src-ip       Source IP address
  src-mac      Source MAC address
SWA(config)#aggregateport load-balance dst-mac           #配置基于目的MAC地址流量平衡
```

（4）在SWB交换机上的端口聚合上配置流量平衡

SWB(config)#aggregateport load-balance dst-mac　　　　#配置基于目的MAC地址流量平衡

任务2：配置三层端口聚合

任务实施

1. 任务描述及网络拓扑设计

在主校区SW1和SW2两台三层交换机之间的冗余链路上实现三层端口聚合，以增加网络骨干链路的带宽，且实现基于源IP地址的流量负载均衡。PC1和PC2分别属于VLAN 10和VLAN 30。绘制拓扑结构图，如图1-5所示。

图1-5　三层端口聚合

2. 网络设备配置

（1）在SW1交换机上创建VLAN、配置SVI

```
SW1(config)#vlan 10
SW1(config-vlan)#exit
SW1(config)#interface fastEthernet 0/10
SW1(config-FastEthernet 0/10)#switchport access vlan 10
SW1(config-FastEthernet 0/10)#exit
SW1(config)#interface vlan 10
SW1(config-VLAN 10)#ip address 192.168.10.1 255.255.255.0
SW1(config-VLAN 10)#exit
```

（2）在SW1交换机上创建三层端口聚合

```
SW1(config)#interface range fastEthernet 0/23-24          #进入一组接口
SW1(config-if-range)#no switchport                        #启用三层功能
SW1(config-if-range)#exit
SW1(config)#interface aggregateport 1                     #进入端口聚合模式
SW1(config-AggregatePort 1)#no switchport                 #启用三层功能
SW1(config-AggregatePort 1)#ip address 10.1.1.1 255.255.255.252
SW1(config-AggregatePort 1)#no shutdown
SW1(config-AggregatePort 1)#exit
SW1(config)#interface range fastEthernet 0/23-24          #进入一组接口
SW1(config-if-range)#port-group 1                         #将接口配置成AP的成员端口
SW1(config-AggregatePort 1)#exit
```

（3）在SW2交换机上创建VLAN、配置SVI

```
SW2(config)#interface fastEthernet 0/10
SW2(config-FastEthernet 0/10)#switchport access vlan 30
SW2(config-FastEthernet 0/10)#exit
SW2(config)#interface vlan 30
SW2(config-VLAN 30)#ip address 192.168.30.1 255.255.255.0
```

（4）在SW2交换机上创建三层端口聚合
SW2(config)#interface range fastEthernet 0/23-24
SW2(config-if-range)#no switchport
SW2(config-if-range)#exit
SW2(config)#interface aggregateport 1
SW2(config-AggregatePort 1)#no switchport
SW2(config-AggregatePort 1)#ip address 10.1.1.2 255.255.255.252
SW2(config-AggregatePort 1)#no shutdown
SW2(config-AggregatePort 1)#exit
SW2(config)#interface range fastEthernet 0/23-24
SW2(config-if-range)#port-group 1

（5）在SW1交换机上配置RIP V2动态路由
SW1(config)#router rip   #启用RIP路由功能
SW1(config-router)#version 2   #设置版本为2
SW1(config-router)#network 192.168.10.0   #宣告直连路由
SW1(config-router)#network 10.1.1.0   #宣告直连路由
SW1(config-router)#no auto-summary   #关闭自动汇总

（6）在SW2交换机上配置RIP V2动态路由
SW2(config)#router rip   #启用RIP路由功能
SW2(config-router)#version 2   #设置版本为2
SW2(config-router)#network 10.1.1.0   #宣告直连路由
SW2(config-router)#network 192.168.30.0   #宣告直连路由
SW2(config-router)#no auto-summary   #关闭自动汇总

（7）验证测试
SW1#show ip route

Codes: C - connected, S - static, R - RIP, B - BGP
        O - OSPF, IA - OSPF inter area
        N1 - OSPF NSSA external type 1, N2 - OSPF NSSA external type 2
        E1 - OSPF external type 1, E2 - OSPF external type 2
        i - IS-IS, su - IS-IS summary, L1 - IS-IS level-1, L2 - IS-IS level-2
        ia - IS-IS inter area, * - candidate default

Gateway of last resort is no set
C   10.1.1.0/30 is directly connected, AggregatePort 1
C   10.1.1.1/32 is local host.
C   192.168.10.0/24 is directly connected, VLAN 10
C   192.168.10.1/32 is local host.
R   192.168.30.0/24 [120/1] via 10.1.1.2, 00:00:04, AggregatePort 1

（8）在SW1交换机端口聚合上配置流量负载均衡
SW1(config)#aggregateport load-balance src-ip   #配置基于源地址流量平衡

（9）在SW2交换机端口聚合上配置流量负载均衡
SW2(config)#aggregateport load-balance src-ip   #配置基于源地址流量平衡

（10）查看端口聚合的配置
1）在SW1交换机查看端口聚合流量负载平衡方式。
SW1#show aggregatePort load-balance
Load-balance   : Source IP

2）在SW1交换机查看端口聚合成员信息。
SW1#show aggregatePort summary
AggregatePort   MaxPorts   SwitchPort Mode   Ports
-------------- -------- ---------- ------ ----------------------------------
Ag1   8   Disabled   Fa0/23，Fa0/24

从show命令输出结果可以看出，端口聚合1包含Fa0/23-24端口，最多可以有8个端口聚合。

3）在SW1交换机查看端口聚合1的状态。
SW1#show interfaces aggregateport 1
Index(dec):27 (hex):1b
AggregatePort 1 is UP , line protocol is UP
Hardware is Aggregate Link AggregatePort, address is 001a.a9bc.7ca3 (bia 001a.a9bc.7ca3)
Interface address is: 10.1.1.1/30
ARP type: ARPA, ARP Timeout: 3600 seconds
　MTU 1500 bytes, BW 200000 Kbit
　Encapsulation protocol is Ethernet-II, loopback not set
　Keepalive interval is 10 sec , set
　Carrier delay is 2 sec
　RXload is 1 ,Txload is 1
　Queueing strategy: FIFO
　　Output queue 0/0, 0 drops;
　　Input queue 0/75, 0 drops
　Switchport attributes:
　　interface's description:""
　　medium-type is copper
　　lastchange time:0 Day: 0 Hour: 3 Minute: 5 Second
　　Priority is 0
　　admin duplex mode is AUTO, oper duplex is Full
　　admin speed is AUTO, oper speed is 100M
　　flow control admin status is OFF,flow control oper status is OFF
　　broadcast Storm Control is ON,multicast Storm Control is OFF,unicast Storm Control is ON
　Aggregate Port Informations:
　　　　Aggregate Number: 1
　　　　Name: "AggregatePort 1"
　　　　Refs: 2
　　　　Members: (count=2)
　　　　　FastEthernet 0/23 Link Status: Up
　　　　　FastEthernet 0/24 Link Status: Up

（11）验证测试

在SW1交换机上长时间ping SW2交换机上VLAN 30地址，然后断开端口聚合中的F0/23端口，观察效果。
SW1#ping 192.168.30.1 ntimes 200
Sending 200, 100-byte ICMP Echoes to 192.168.30.1, timeout is 2 seconds:
  < press Ctrl+C to break >
!!!
!!.!!!!!!!!!!!!!!!!!!!!!!!!!!!!!

第1章 交换技术

```
!!!!!!!!!!!!!!!!!!!!!!!!!!!!!!!!!!!!!!!!!
Success rate is 99 percent (199/200), round-trip min/avg/max = 1/1/10 ms
```

此时发现有一个丢包。说明在负载均衡方式下，同一对源地址和目的地址之间的流量只从一个物理端口转发，一个端口断开时会将流量切换到另一个端口上，引起链路短暂中断的现象。

小结

通过端口聚合技术的学习，主要掌握二层端口聚合、三层端口聚合以及基于端口聚合的流量负载均衡等配置与管理工作。

默认情况下，一个端口聚合是一个二层的AP。如果要配置一个三层AP，则需要使用no switchport命令将其设置为三层端口。

1.3 生成树技术

问题描述

随着该校各个部门对网络的依赖越来越强，为了保证网络的高可用性，有时会希望在网络中提供设备、模块和链路的冗余。

分校区采用双核心网络架构，为了保障二层链路的冗余和负载均衡，需要配置MSTP协议。创建两个MSTP实例，分别为实例1和实例2。实例1的成员包含VLAN 40和VLAN 50，实例2的成员包含VLAN 60，设置两台三层交换机为生成树实例的根，并要求两台三层交换机互为备份根。

为了提高网络的可靠性，在接入层交换机和核心交换机上用两条链路实现互联，以增加网络骨干链路的带宽，运行STP/RSTP技术，使网络避免环路。

问题分析

生成树技术主要涉及STP协议配置、RSTP协议配置以及MSTP协议配置。

1. 配置STP/RSTP

步骤1 开启生成树协议。

switch(config)#**spanning-tree**

步骤2 配置生成树模式，可以根据需要选择生成树版本是STP或者RSTP。

switch(config)#spanning-tree mode { stp | rstp }

步骤3 配置交换机的优先级，优先级是4096的倍数，默认值是32768。

switch(config)#spanning-tree priority *<0-61440>*

步骤4 配置端口优先级，端口优先级是16的倍数，默认值是128。

switch(config-if)#spanning-tree priority *<0-240>*

步骤5 配置端口的路径成本。（可选）

switch(config-if)#**spanning-tree cost** *cost*

2. MSTP基本配置

步骤1 开启生成树协议。

switch(config)#**spanning-tree**

步骤2 配置生成树模式为MSTP。

switch(config)#spanning-tree mode mstp

步骤3 进入MSTP配置模式。

switch(config)#spanning-tree mst configuration

步骤4 配置VLAN与生成树实例的映射关系。instance-id是实例名称。

switch(config-mst)#**instance** *instance-id* **vlan** *vlan-range*

步骤5 配置MST区域的配置名称，参数name表示区域的名称。

switch(config-mst)#**name** *name*

步骤6 配置MST区域的修正号，参数的取值范围是0～65535，默认值为0。

switch(config-mst)#**revision** *number*

任 务 单

1	STP/RSTP技术配置
2	MSTP技术配置

解决步骤

根据任务单的安排完成任务。

任务1：STP/RSTP技术配置

任务实施

1. 任务描述及网络拓扑设计

为了提高网络的可靠性，在SWA和SWB交换机上用两条链路实现互联，以增加网络骨干链路的带宽，运行STP/RSTP技术，使网络避免环路。绘制拓扑结构图，如图1-6所示。

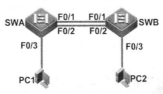

图1-6　生成树协议

2. 网络设备配置

（1）完成Trunk配置

1）在SWA交换机上完成Trunk配置。

SWA(config)#interface range fastEthernet 0/1-2

SWA(config-if-range)#switchport mode trunk
SWA(config-if-range)#exit
2）在SWB交换机上完成Trunk配置。
SWB(config)#interface range fastEthernet 0/1-2
SWB(config-if-range)#switchport mode trunk
SWB(config-if-range)#exit

（2）配置快速生成树协议，并且指定SWA为根交换机
1）在SWA交换机上配置生成树协议。
SWA(config)#spanning-tree   #开启生成树协议
SWA(config)#spanning-tree mode rstp   #配置生成树模式为RSTP
SWA(config)#spanning-tree priority 4096
　　　　　　#指定交换机优先级为4096，默认为32768，指定SWA为根交换机
2）在SWB交换机上配置生成树协议。
SWB(config)#spanning-tree   #开启生成树协议
SWB(config)#spanning-tree mode rstp   #配置生成树模式为RSTP
SWB(config)#

（3）查看交换机及端口的状态
SWB#show spanning-tree   #查看生成树协议
StpVersion : RSTP   #协议版本类型为RSTP
SysStpStatus : ENABLED   #协议为开启状态
MaxAge : 20
HelloTime : 2
ForwardDelay : 15
BridgeMaxAge : 20
BridgeHelloTime : 2
BridgeForwardDelay : 15
MaxHops: 20
TxHoldCount : 3
PathCostMethod : Long
BPDUGuard : Disabled
BPDUFilter : Disabled
LoopGuardDef : Disabled
BridgeAddr : 001a.a97f.ef11
Priority: 32768
TimeSinceTopologyChange : 0d:0h:0m:17s
TopologyChanges : 2
DesignatedRoot : 1000.001a.a9bc.7ca2
RootCost : 200000   #交换机到根交换机的开销为200000
RootPort : 1   #根端口为F0/1
从show命令的输出结果看以看到交换机SWB为非根交换机，根端口为F0/1。
SWB#show spanning-tree interface fastEthernet 0/1   #查看端口RSTP状态
PortAdminPortFast : Disabled
PortOperPortFast : Disabled
PortAdminAutoEdge : Enabled
PortOperAutoEdge : Disabled
PortAdminLinkType : auto
PortOperLinkType : point-to-point
PortBPDUGuard : Disabled

PortBPDUFilter : Disabled
　　PortGuardmode : None
　　PortState : forwarding　　　　　　　　　　　　　#F0/1端口处于转发状态
　　PortPriority : 128
　　PortDesignatedRoot : 1000.001a.a9bc.7ca2
　　PortDesignatedCost : 0
　　PortDesignatedBridge :1000.001a.a9bc.7ca2
　　PortDesignatedPort : 8001
　　PortForwardTransitions : 1
　　PortAdminPathCost : 200000
　　PortOperPathCost : 200000
　　Inconsistent states : normal
　　PortRole : rootPort　　　　　　　　　　　　　　#显示端口角色为根端口

从show命令的输出结果看以看到交换机SWB的端口为F0/1，角色为根端口，处于转发状态。

　　SWB#show spanning-tree interface fastEthernet 0/2
　　PortAdminPortFast : Disabled
　　PortOperPortFast : Disabled
　　PortAdminAutoEdge : Enabled
　　PortOperAutoEdge : Disabled
　　PortAdminLinkType : auto
　　PortOperLinkType : point-to-point
　　PortBPDUGuard : Disabled
　　PortBPDUFilter : Disabled
　　PortGuardmode : None
　　PortState : discarding　　　　　　　　　　　　　#F0/2端口处于阻塞状态
　　PortPriority : 128
　　PortDesignatedRoot : 1000.001a.a9bc.7ca2
　　PortDesignatedCost : 0
　　PortDesignatedBridge :1000.001a.a9bc.7ca2
　　PortDesignatedPort : 8002
　　PortForwardTransitions : 2
　　PortAdminPathCost : 200000
　　PortOperPathCost : 200000
　　Inconsistent states : normal
　　PortRole : alternatePort　　　　　　　　　　　　#F0/2端口为根端口的替换端口

从show命令的输出结果看以看到交换机SWB的端口为F0/2，角色为替换端口，状态为阻塞状态。

　　（4）设置SWB交换机的F0/2端口为根端口
　　SWB(config)#interface fastEthernet 0/2
　　SWB(config-FastEthernet 0/2)#spanning-tree port-priority 16
　　　　　#设定F0/2端口优先级为16，默认为128，该端口所连接链路为主链路
　　（5）查看交换机及端口的状态
　　SWA#show spanning-tree　　　　　　　　　　　　#查看生成树协议
　　StpVersion : RSTP
　　SysStpStatus : ENABLED
　　MaxAge : 20
　　HelloTime : 2

第1章 交换技术

```
ForwardDelay : 15
BridgeMaxAge : 20
BridgeHelloTime : 2
BridgeForwardDelay : 15
MaxHops: 20
TxHoldCount : 3
PathCostMethod : Long
BPDUGuard : Disabled
BPDUFilter : Disabled
LoopGuardDef : Disabled
BridgeAddr : 001a.a9bc.7ca2
Priority: 4096
TimeSinceTopologyChange : 0d:0h:0m:5s
TopologyChanges : 12
DesignatedRoot : 1000.001a.a9bc.7ca2
RootCost : 0
RootPort : 0                                    #表示SWA没有根端口,为根交换机
```

上述show命令输出结果显示交换机SWA为根交换机。

```
SWB#show spanning-tree
StpVersion : RSTP
SysStpStatus : ENABLED
MaxAge : 20
HelloTime : 2
ForwardDelay : 15
BridgeMaxAge : 20
BridgeHelloTime : 2
BridgeForwardDelay : 15
MaxHops: 20
TxHoldCount : 3
PathCostMethod : Long
BPDUGuard : Disabled
BPDUFilter : Disabled
LoopGuardDef : Disabled
BridgeAddr : 001a.a97f.ef11
Priority: 32768
TimeSinceTopologyChange : 0d:0h:0m:48s
TopologyChanges : 6
DesignatedRoot : 1000.001a.a9bc.7ca2
RootCost : 200000
RootPort : 2                                    #显示SWB的根端口为F0/2
```

上述show命令输出结果显示交换机SWB的F0/2端口为根端口,处于转发状态。

(6) 验证测试

按图1-6连接拓扑,给PC1主机配置IP地址为192.168.1.10/24,给PC2主机配置IP地址为192.168.1.20/24。

1) 让交换机SWA与SWB之间的F0/1端口所连接备用链路断掉(如拔掉网线),观察PC1 ping PC2效果。

```
C:\Documents and Settings\Administrator>ping 192.168.1.20
```

Pinging 192.168.1.20 with 32 bytes of data:

Reply from 192.168.1.20: bytes=32 time<1ms TTL=64
Reply from 192.168.1.20: bytes=32 time<1ms TTL=64
Reply from 192.168.1.20: bytes=32 time<1ms TTL=64
Reply from 192.168.1.20: bytes=32 time<1ms TTL=64

Ping statistics for 192.168.1.20:
    Packets: Sent = 4, Received = 4, Lost = 0 (0% loss),
Approximate round trip times in milli-seconds:
    Minimum = 0ms, Maximum = 0ms, Average = 0ms

2）重新接上F0/1端口所连接链路，待网络稳定后，让交换机SWA与SWB之间的F0/2端口所连接主链路断掉，观察PC1 ping PC2效果。
C:\Documents and Settings\Administrator>ping 192.168.1.20 -t

Pinging 192.168.1.20 with 32 bytes of data:

Reply from 192.168.1.20: bytes=32 time<1ms TTL=64
Reply from 192.168.1.20: bytes=32 time<1ms TTL=64
Reply from 192.168.1.20: bytes=32 time<1ms TTL=64
Reply from 192.168.1.20: bytes=32 time<1ms TTL=64
Request timed out.
Reply from 192.168.1.20: bytes=32 time<1ms TTL=64
Reply from 192.168.1.20: bytes=32 time<1ms TTL=64
Reply from 192.168.1.20: bytes=32 time<1ms TTL=64
Reply from 192.168.1.20: bytes=32 time<1ms TTL=64

以上结果显示丢包数为一个。

任务2：MSTP技术配置

任务实施

1. 任务描述及网络拓扑设计

利用MSTP除了可以实现网络中的冗余链路外，还能够在实现网络冗余和可靠性的同时实现负载均衡。在图1-7所示拓扑结构中，PC1和PC3在VLAN 40中，PC2在VLAN 50中，PC4在VLAN 60中。

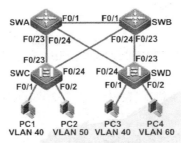

图1-7　多生成树协议

2. 网络设备配置

（1）在SWA、SWB、SWC、SWD设备上完成VLAN划分及Trunk配置

1）在SWA交换机上创建VLAN、配置Trunk。
SWA(config)#vlan 40
SWA(config-vlan)#exit
SWA(config)#vlan 50
SWA(config-vlan)#exit
SWA(config)#vlan 60
SWA(config-vlan)#exit
SWA(config)#interface fastEthernet 0/1
SWA(config-FastEthernet 0/1)#switchport mode trunk
SWA(config-FastEthernet 0/1)#exit
SWA(config)#interface fastEthernet 0/23
SWA(config-FastEthernet 0/23)#switchport mode trunk
SWA(config-FastEthernet 0/23)#exit
SWA(config)#interface fastEthernet 0/24
SWA(config-FastEthernet 0/24)#switchport mode trunk
SWA(config-FastEthernet 0/24)#exit

2）在SWB交换机上创建VLAN、配置Trunk。
SWB(config)#vlan 40
SWB(config-vlan)#exit
SWB(config)#vlan 50
SWB(config-vlan)#exit
SWB(config)#vlan 60
SWB(config-vlan)#exit
SWB(config)#interface fastEthernet 0/1
SWB(config-FastEthernet 0/1)#switchport mode trunk
SWB(config-FastEthernet 0/1)#exit
SWB(config)#interface fastEthernet 0/23
SWB(config-FastEthernet 0/23)#switchport mode trunk
SWB(config-FastEthernet 0/23)#exit
SWB(config)#interface fastEthernet 0/24
SWB(config-FastEthernet 0/24)#switchport mode trunk
SWB(config-FastEthernet 0/24)#exit

3）在SWC交换机上创建VLAN、配置Trunk。
SWC(config)#vlan 40
SWC(config-vlan)#exit
SWC(config)#vlan 50
SWC(config-vlan)#exit
SWC(config)#vlan 60
SWC(config-vlan)#exit
SWC(config)#interface fastEthernet 0/1
SWC(config-FastEthernet 0/1)#switchport access vlan 40
SWC(config-FastEthernet 0/1)#exit
SWC(config)#interface fastEthernet 0/2
SWC(config-FastEthernet 0/2)# switchport access vlan 50
SWC(config-FastEthernet 0/2)#exit
SWC(config)#interface fastEthernet 0/23
SWC(config-FastEthernet 0/23)#switchport mode trunk
SWC(config-FastEthernet 0/23)#exit
SWC(config)#interface fastEthernet 0/24

SWC(config-FastEthernet 0/24)#switchport mode trunk
SWC(config-FastEthernet 0/24)#exit

4）在SWD交换机上创建VLAN、配置Trunk。
SWD(config)#vlan 40
SWD(config-vlan)#exit
SWD(config)#vlan 50
SWD(config-vlan)#exit
SWD(config)#vlan 60
SWD(config-vlan)#exit
SWD(config)#interface fastEthernet 0/1
SWD(config-FastEthernet 0/1)#switchport access vlan 40
SWD(config-FastEthernet 0/1)#exit
SWD(config)#interface fastEthernet 0/2
SWD(config-FastEthernet 0/2)# switchport access vlan 60
SWD(config-FastEthernet 0/2)#exit
SWD(config)#interface fastEthernet 0/23
SWD(config-FastEthernet 0/23)#switchport mode trunk
SWD(config-FastEthernet 0/23)#exit
SWD(config)#interface fastEthernet 0/24
SWD(config-FastEthernet 0/24)#switchport mode trunk
SWD(config-FastEthernet 0/24)#exit

（2）在SWA、SWB、SWC、SWD设备上配置快速生成树协议

1）在SWA交换机上配置生成树协议。
SWA(config)#spanning-tree   #开启生成树协议
SWA(config)#spanning-tree mode mstp   #配置协议类型为MSTP

2）在SWB交换机上配置生成树协议。
SWB(config)#spanning-tree   #开启生成树协议
SWB(config)#spanning-tree mode mstp   #配置协议类型为MSTP

3）在SWC交换机上配置生成树协议。
SWC(config)#spanning-tree   #开启生成树协议
SWC(config)#spanning-tree mode mstp   #配置协议类型为MSTP

4）在SWD交换机上配置生成树协议。
SWD(config)#spanning-tree
SWD(config)#spanning-tree mode mstp

（3）在SWA、SWB、SWC、SWD设备上配置MSTP

1）在SWA交换机上配置MSTP。
SWA(config)#spanning-tree mst 1 priority 4096
#配置交换机在实例1中的优先级为4096，使其在instance 1中成为根
SWA(config)#spanning-tree mst configuration   #进入MSTP配置模式
SWA(config-mst)#instance 1 vlan 1,40   #配置实例1关联vlan 1和vlan 40
SWA(config-mst)#instance 2 vlan 50,60   #配置实例2关联vlan 50和vlan 60
SWA(config-mst)#name region1   #配置域名称
SWA(config-mst)#revision 1   #配置修订号
SWA(config-mst)#exit

2）在SWB交换机上配置MSTP。
SWB(config)#spanning-tree mst 2 priority 4096
#配置交换机在实例2中的优先级为4096，使其在instance 2中成为根

```
SWB(config)#spanning-tree mst configuration          #进入MSTP配置模式
SWB(config-mst)#instance 1 vlan 1,40                 #配置实例1关联vlan 1和vlan 40
SWB(config-mst)#instance 2 vlan 50,60                #配置实例2关联vlan 50和vlan 60
SWB(config-mst)#name region1                         #配置域名称
SWB(config-mst)#revision 1                           #配置修订号
SWB(config-mst)#exit
```

3)在SWC交换机上配置MSTP。
```
SWC(config)#spanning-tree mst configuration          #进入MSTP配置模式
SWC(config-mst)#instance 1 vlan 1,40                 #配置实例1关联vlan 1和vlan 40
SWC(config-mst)#instance 2 vlan 50,60                #配置实例2关联vlan 50和vlan 60
SWC(config-mst)#name region1                         #配置域名称
SWC(config-mst)#revision 1                           #配置修订号
SWC(config-mst)#exit
```

4)在SWD交换机上配置MSTP。
```
SWD(config)#spanning-tree mst configuration          #进入MSTP配置模式
SWD(config-mst)#instance 1 vlan 1,40                 #配置实例1关联vlan 1和vlan 40
SWD(config-mst)#instance 2 vlan 50,60                #配置实例2关联vlan 50和vlan 60
SWD(config-mst)#name region1                         #配置域名称
SWD(config-mst)#revision 1                           #配置修订号
SWD(config-mst)#exit
```

(4)查看交换机MSTP选举结果

1)在SWA交换机上查看MSTP选举结果。
```
SWA#show spanning-tree mst 1
MST 1 vlans mapped : 1, 40
BridgeAddr : 001a.a9bc.7ca2
Priority: 4096
TimeSinceTopologyChange : 0d:0h:1m:10s
TopologyChanges : 17
DesignatedRoot : 1001.001a.a9bc.7ca2
RootCost : 0
RootPort : 0
```
从上述show命令输出结果可以看出交换机SWA为实例1中的根交换机。

2)在SWB交换机上查看MSTP选举结果。
```
SWB#show spanning-tree mst 2
MST 2 vlans mapped : 50, 60
BridgeAddr : 001a.a97f.ef11
Priority: 4096
TimeSinceTopologyChange : 0d:0h:2m:35s
TopologyChanges : 6
DesignatedRoot : 1002.001a.a97f.ef11
RootCost : 0
RootPort : 0
```
从上述show命令输出结果可以看出交换机SWB为实例2中的根交换机。

3)在SWC交换机上查看实例1中MSTP选举结果。
```
SWC#show spanning-tree mst 1
MST 1 vlans mapped : 1,40
BridgeAddr : 00d0.f8ff.4728
Priority : 32768
```

TimeSinceTopologyChange : 0d:1h:28m:27s
TopologyChanges : 0
DesignatedRoot : 1001001AA9BC7CA2
RootCost : 200000
RootPort : Fa0/23

从上述show命令输出结果可以看出，在实例1中，交换机SWC的端口F0/23为根端口，因此VLAN1和VLAN 10的数据经端口F0/23转发。

4）在SWC交换机上查看实例2中MSTP的选举结果。
SWC#show spanning-tree mst 2
MST 2 vlans mapped : 50,60
BridgeAddr : 00d0.f8ff.4728
Priority : 32768
TimeSinceTopologyChange : 0d:1h:28m:33s
TopologyChanges : 0
DesignatedRoot : 1002001AA97FEF11
RootCost : 200000
RootPort : Fa0/24

从上述show命令输出结果可以看出，在实例2中，交换机SWC的端口F0/24为根端口，因此VLAN20和VLAN 40的数据经端口F0/24转发。

小结

通过生成树技术的学习，主要掌握STP、RSTP和MSTP协议等配置与管理工作。

生成树协议通过逻辑上阻塞一些冗余端口来消除环路，将物理环路改变为逻辑上无环路的拓扑，而一旦活动链路发生故障，被阻塞的端口能够立即启用，以达到冗余备份的目的。

MSTP引入了MST区域和实例的概念。在MSTP中，每个实例都将计算出一个独立的生成树。可以将一个或多个VLAN映射到一个实例中，这样不同的VLAN之间将存在不同的选举结果，从而避免了连通性丢失的问题，并起到对流量负载分担的作用。

第2章 路由技术

路由器提供了在异构网络互连机制中，实现将数据包从一个网络发送到另一个网络的功能。路由就是指导IP数据包发送的路径信息。

路由器转发数据包的关键是路由表。每个路由器中都保存着一张路由表，表中每条路由项都指明数据到某个子网应通过路由器的哪个物理接口发送出去。在互联网中进行路由选择时，路由器只是根据所收到的数据包的目的地址选择一个合适的路径，将数据包传送到下一跳路由器，路径上最后的路由器负责将数据包送交目的主机。每个路由器只负责自己本站数据包通过最优的路径转发，通过多个路由器一站一站地将数据包通过最佳路径转发到目的地。当然有时候由于实施一些路由策略，以超越传统的路由选择方式自由地实现报文转发和路径选择，数据包通过的路径并不一定是最佳路由。

路由表的产生方式一般有直连路由、静态路由、动态路由3种。

1) 直连路由：给路由器接口配置一个IP地址，路由器自动产生本接口IP所在网段的路由信息。

2) 静态路由：在拓扑结构简单的网络中，网络管理员通过手工方式配置本路由器未知网段的路由信息，从而实现不同网段之间的连接。

3) 动态路由：在大规模的网络中，通过在路由器上运行动态路由协议，路由器之间通过互相自动学习产生路由信息。

本章介绍的路由技术主要涉及静态路由、RIP路由、OSPF路由、路由重分发及路由控制、策略路由。

1. 静态路由

静态路由是指网络管理员手工配置的路由信息。它是一种最简单的配置路由的方法，一般用在小型网络或拓扑相对固定的网络中。

2. RIP 路由

RIP协议是基于距离矢量算法的一种路由协议。它使用条数来衡量到达目标地址的路由距离。启用RIP V2的路由器更新方式为组播更新，即以224.0.0.9为目的地址发送更新，这样做可以尽量减少由于更新引起的广播流量，增加链路带宽的使用效率。

3. OSPF路由

OSPF是采用链路状态技术，路由器互相发送直接相连的链路信息和它所拥有的到其他路由器的链路信息。每个OSPF路由器维护相同自治系统拓扑结构的数据库。从这个数据库里，可以构造出最短路径树来计算出路由表。当拓扑结构发生变化时，OSPF能够迅速重新计算出路径，而只产生少量的路由协议流量。

4. 路由重分发及路由控制

路由重分发是指连接到不同路由选择域的边界路由器，在不同路由选择域之间交换和通告路由选择信息的能力。

路由过滤是指对进出站路由进行过滤，使得路由器只学到必要的、可预知的路由，对外只向可信任的路由器通告必要的、可预知的路由。

5. 策略路由

策略路由是一种入站机制，用于入站报文。通过使用基于策略的路由选择，能够根据数据包的源地址、目的地址、源端口、目的端口和协议类型让报文选择不同的路径。

使用策略路由可以灵活地操纵报文，将不同情况的数据流通过不同的路径传输，这样就提高了服务质量，满足了系统及用户的需求，达到了负载均衡的效果。当使用策略路由时，路由器对于接收的报文将不进行传统的查找路由表的工作，而是根据配置的策略将数据发送到相应的路径。

2.1 静态路由

问题描述

学校分校区网络规模较小，网络架构采用二层架构，在路由规划时，可采用静态路由，在互联网络的路由选择行为上实施非常精确的控制。

问题分析

静态路由是由管理员手工配置的路由信息。默认路由是指路由表中未直接列出目标网络选项的路由选择项，它用于在不明确的情况下指示数据帧下一跳的方向，它可以看作是静态路由的一种特殊情况。路由器如果配置了默认路由，则所有未明确指明目标网络的数据包都按默认路由进行转发。浮动静态路由不能被永久地保存在路由选择表中，它仅仅会出现在一条首选路由发生失败的时候，它主要考虑到链路的冗余性能。

1. 配置静态路由

ip route *network-number network-mask* {*ip-address* | *interface-id*}
#参数network-number表示目标网段，network-mask表示目标网段子网掩码，ip-address表示下一跳IP地址，interface-id表示接口号

2. 配置默认路由

ip route 0.0.0.0 0.0.0.0 {*ip-address* | *interface-id*}
#参数ip-address表示下一跳IP地址，interface-id表示接口号

3. 配置浮动静态路由

ip route network-number network-mask {ip-address | interface-id} distance

第2章 路由技术

#参数network-number表示目标网段，network-mask表示目标网段子网掩码，ip-address表示下一跳IP地址，interface-id表示接口号，distance表示管理距离。

任 务 单

1	静态路由配置
2	浮动静态路由配置

根据任务单的安排完成任务。

任务1：静态路由配置

任务实施

1. 任务描述及网络拓扑设计

在RA、RB与RC路由器上配置静态路由，实现全网互通。绘制拓扑结构图，如图2-1所示。

图2-1 静态路由

2. 网络设备配置

（1）配置路由器各接口的IP地址

1）在RA路由器上配置IP地址。
RA(config)#interface fastEthernet 0/0
RA(config-if-FastEthernet 0/0)#ip address 192.168.0.1 255.255.255.0
RA(config-if-FastEthernet 0/0)#no shutdown
RA(config-if-FastEthernet 0/0)#exit
RA(config)#interface fastEthernet 0/1
RA(config-if-FastEthernet 0/1)#ip address 192.168.1.1 255.255.255.0
RA(config-if-FastEthernet 0/1)#no shutdown
RA(config-if-FastEthernet 0/1)#exit

2）在RB路由器上配置IP地址。
RB(config)#interface fastEthernet 0/0
RB(config-if-FastEthernet 0/0)#ip address 192.168.2.1 255.255.255.0
RB(config-if-FastEthernet 0/0)#no shutdown
RB(config-if-FastEthernet 0/0)#exit
RB(config)#interface fastEthernet 0/1
RB(config-if-FastEthernet 0/1)#ip address 192.168.1.2 255.255.255.0
RB(config-if-FastEthernet 0/1)#no shutdown
RB(config-if-FastEthernet 0/1)#exit

3）在RC路由器上配置IP地址。
RC(config)#interface fastEthernet 0/0
RC(config-if-FastEthernet 0/0)#ip address 192.168.2.2 255.255.255.0
RC(config-if-FastEthernet 0/0)#no shutdown
RC(config-if-FastEthernet 0/0)#exit
RC(config)#interface fastEthernet 0/1
RC(config-if-FastEthernet 0/1)#ip address 192.168.3.1 255.255.255.0
RC(config-if-FastEthernet 0/1)#no shutdown
RC(config-if-FastEthernet 0/1)#exit

（2）配置静态路由
1）在RA路由器上配置静态路由。
RA(config)#ip route 192.168.2.0 255.255.255.0 192.168.1.2   #配置到非直连网段静态路由
RA(config)#ip route 192.168.3.0 255.255.255.0 192.168.1.2   #配置到非直连网段静态路由
2）在RB路由器上配置静态路由。
RB(config)#ip route 192.168.0.0 255.255.255.0 192.168.1.1   #配置到非直连网段静态路由
RB(config)#ip route 192.168.3.0 255.255.255.0 192.168.2.2   #配置到非直连网段静态路由
3）在RC路由器上配置静态路由。
RC(config)#ip route 192.168.0.0 255.255.255.0 192.168.2.1   #配置到非直连网段静态路由
RC(config)#ip route 192.168.1.0 255.255.255.0 192.168.2.1   #配置到非直连网段静态路由

（3）验证测试
1）查看路由表。
RA#show ip route

Codes: C - connected, S - static, R - RIP, B - BGP
         O - OSPF, IA - OSPF inter area
         N1 - OSPF NSSA external type 1, N2 - OSPF NSSA external type 2
         E1 - OSPF external type 1, E2 - OSPF external type 2
         i - IS-IS, su - IS-IS summary, L1 - IS-IS level-1, L2 - IS-IS level-2
         ia - IS-IS inter area, * - candidate default

Gateway of last resort is no set
C   192.168.0.0/24 is directly connected, FastEthernet 0/0
C   192.168.0.1/32 is local host.
C   192.168.1.0/24 is directly connected, FastEthernet 0/1
C   192.168.1.1/32 is local host.
S   192.168.2.0/24 [1/0] via 192.168.1.2
S   192.168.3.0/24 [1/0] via 192.168.1.2

从查看路由表的输出结果可以看到，192.168.2.0/24和192.168.3.0/24网段是通过下一跳地址获得的。

2）测试网络连通性。
RA#ping 192.168.0.1
  < press Ctrl+C to break >
!!!!!
 Success rate is 100 percent (5/5), round-trip min/avg/max = 1/2/10 ms
RA#ping 192.168.1.1
Sending 5, 100-byte ICMP Echoes to 192.168.1.1, timeout is 2 seconds:
  < press Ctrl+C to break >

第2章 路由技术

!!!!!
Success rate is 100 percent (5/5), round-trip min/avg/max = 1/1/1 ms
RA#ping 192.168.2.1
Sending 5, 100-byte ICMP Echoes to 192.168.2.1, timeout is 2 seconds:
　< press Ctrl+C to break >
!!!!!
Success rate is 100 percent (5/5), round-trip min/avg/max = 1/2/10 ms
RA#ping 192.168.2.2
Sending 5, 100-byte ICMP Echoes to 192.168.2.2, timeout is 2 seconds:
　< press Ctrl+C to break >
!!!!!
Success rate is 100 percent (5/5), round-trip min/avg/max = 1/2/10 ms
RA#ping 192.168.3.1
Sending 5, 100-byte ICMP Echoes to 192.168.3.1, timeout is 2 seconds:
　< press Ctrl+C to break >
!!!!!
Success rate is 100 percent (5/5), round-trip min/avg/max = 1/2/10 ms

任务2：浮动静态路由配置

任务实施

1. 任务描述及网络拓扑设计

RA路由器去往RD路由器的网络有两条路径，其首选的路径是RA-RB-RD，为了保证链路的可用性，需要一条备份链路。在主链路断开的时候，让其从备份链路RA-RC-RD这条路径传输数据，当主链路恢复时，还会用主链路传输数据。绘制拓扑结构图，如图2-2所示。

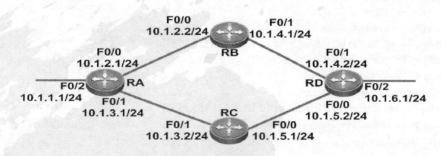

图2-2　浮动静态路由

2. 网络设备配置

（1）配置路由器各接口的IP地址

1）在RA路由器上配置IP地址。
RA(config)#interface fastEthernet 0/0
RA(config-if-FastEthernet 0/0)#ip address 10.1.2.1 255.255.255.0
RA(config-if-FastEthernet 0/0)#no shutdown
RA(config-if-FastEthernet 0/0)#exit
RA(config)#interface fastEthernet 0/1

RA(config-if-FastEthernet 0/1)#ip address 10.1.3.1 255.255.255.0
RA(config-if-FastEthernet 0/1)#no shutdown
RA(config-if-FastEthernet 0/1)#exit
RA(config)#interface fastEthernet 0/2
RA(config-if-FastEthernet 0/2)#ip address 10.1.1.1 255.255.255.0
RA(config-if-FastEthernet 0/2)#no shutdown

2）在RB路由器上配置IP地址。
RB(config)#interface fastEthernet 0/0
RB(config-if-FastEthernet 0/0)#ip address 10.1.2.2 255.255.255.0
RB(config-if-FastEthernet 0/0)#no shutdown
RB(config-if-FastEthernet 0/0)#exit
RB(config)#interface fastEthernet 0/1
RB(config-if-FastEthernet 0/1)#ip address 10.1.4.1 255.255.255.0
RB(config-if-FastEthernet 0/1)#no shutdown
RB(config-if-FastEthernet 0/1)#exit

3）在RC路由器上配置IP地址。
RC(config)#interface fastEthernet 0/0
RC(config-if-FastEthernet 0/0)#ip address 10.1.5.1 255.255.255.0
RC(config-if-FastEthernet 0/0)#no shutdown
RC(config-if-FastEthernet 0/0)#exit
RC(config)#interface fastEthernet 0/1
RC(config-if-FastEthernet 0/1)#ip address 10.1.3.2 255.255.255.0
RC(config-if-FastEthernet 0/1)#no shutdown
RC(config-if-FastEthernet 0/1)#exit

4）在RD路由器上配置IP地址。
RD(config)#interface fastEthernet 0/0
RD(config-if-FastEthernet 0/0)#ip address 10.1.5.2 255.255.255.0
RD(config-if-FastEthernet 0/0)#no shutdown
RD(config-if-FastEthernet 0/0)#exit
RD(config)#interface fastEthernet 0/1
RD(config-if-FastEthernet 0/1)#ip address 10.1.4.2 255.255.255.0
RD(config-if-FastEthernet 0/1)#no shutdown
RD(config-if-FastEthernet 0/1)#exit
RD(config)#interface fastEthernet 0/2
RD(config-if-FastEthernet 0/2)#ip address 10.1.6.1 255.255.255.0
RD(config-if-FastEthernet 0/2)#no shutdown

（2）配置静态路由
RA(config)#ip route 10.1.4.0 255.255.255.0 10.1.2.2   #配置静态路由
RA(config)#ip route 10.1.5.0 255.255.255.0 10.1.3.2   #配置静态路由
RA(config)#ip route 10.1.6.0 255.255.255.0 10.1.2.2   #配置静态路由
RA(config)#ip route 10.1.6.0 255.255.255.0 10.1.3.2 10   #配置浮动静态路由

（3）验证测试
1）当网络连通的时候，观察RA的路由表。
RA#show ip route

Codes: C - connected, S - static, R - RIP, B - BGP
        O - OSPF, IA - OSPF inter area
        N1 - OSPF NSSA external type 1, N2 - OSPF NSSA external type 2

E1 - OSPF external type 1, E2 - OSPF external type 2
i - IS-IS, su - IS-IS summary, L1 - IS-IS level-1, L2 - IS-IS level-2
ia - IS-IS inter area, * - candidate default

Gateway of last resort is no set
C   10.1.1.0/24 is directly connected, FastEthernet 0/2
C   10.1.1.1/32 is local host.
C   10.1.2.0/24 is directly connected, FastEthernet 0/0
C   10.1.2.1/32 is local host.
C   10.1.3.0/24 is directly connected, FastEthernet 0/1
C   10.1.3.1/32 is local host.
S   10.1.4.0/24 [1/0] via 10.1.2.2
S   10.1.5.0/24 [1/0] via 10.1.3.2
S   10.1.6.0/24 [1/0] via 10.1.2.2

2）当RA接口F0/0端口状态为down，当主链路失效时。
RA#show ip route

Codes: C - connected, S - static, R - RIP, B - BGP
        O - OSPF, IA - OSPF inter area
        N1 - OSPF NSSA external type 1, N2 - OSPF NSSA external type 2
        E1 - OSPF external type 1, E2 - OSPF external type 2
        i - IS-IS, su - IS-IS summary, L1 - IS-IS level-1, L2 - IS-IS level-2
        ia - IS-IS inter area, * - candidate default

Gateway of last resort is no set
C   10.1.1.0/24 is directly connected, FastEthernet 0/2
C   10.1.1.1/32 is local host.
C   10.1.3.0/24 is directly connected, FastEthernet 0/1
C   10.1.3.1/32 is local host.
S   10.1.5.0/24 [1/0] via 10.1.3.2
S   10.1.6.0/24 [10/0] via 10.1.3.2

3）主链路恢复以后，当RA接口F0/0端口状态为up时。
RA#show ip route

Codes: C - connected, S - static, R - RIP, B - BGP
        O - OSPF, IA - OSPF inter area
        N1 - OSPF NSSA external type 1, N2 - OSPF NSSA external type 2
        E1 - OSPF external type 1, E2 - OSPF external type 2
        i - IS-IS, su - IS-IS summary, L1 - IS-IS level-1, L2 - IS-IS level-2
        ia - IS-IS inter area, * - candidate default

Gateway of last resort is no set
C   10.1.1.0/24 is directly connected, FastEthernet 0/2
C   10.1.1.1/32 is local host.
C   10.1.2.0/24 is directly connected, FastEthernet 0/0
C   10.1.2.1/32 is local host.
C   10.1.3.0/24 is directly connected, FastEthernet 0/1
C   10.1.3.1/32 is local host.
S   10.1.4.0/24 [1/0] via 10.1.2.2

S   10.1.5.0/24 [1/0] via 10.1.3.2
S   10.1.6.0/24 [1/0] via 10.1.2.2

测试结果说明，当RA路由器接口F0/0的状态断掉，且主链路的链路失效时，查看路由表发现所有路由的下一跳指向了10.1.3.2。由于原来的首选路由不再可用，所以路由器切换到管理距离为20的备份链路。而且因为子网10.1.2.0发生故障，所以路由表中不再把它作为直连路由。

小结

通过静态路由的学习，主要掌握静态路由、默认路由、浮动静态路由等配置与管理工作。

默认路由一般使用在stub网络中，stub网络是只有一条出口路径的网络。使用默认路由来发送那些目标网络没有包含在路由表中的数据包。

浮动静态路由不能被永久地保存在路由选择表中，它仅仅会出现在一条首选路由发生失败的时候。浮动静态路由主要考虑到链路的冗余性能。

2.2 RIP V2路由

问题描述

学校分校区网络规模较小，网络架构采用二层架构，在路由规划时，需要选择适合于小型网络的动态路由协议，可采用动态路由协议RIP。为了节省IP地址，需要选择支持VLSM的动态路由协议，RIP V2是无类路由协议，支持VLSM。为保障路由更新的安全性，RIP路由协议需要进行基于接口的MD5验证。

问题分析

在配置RIP路由协议时，如果不配置路由协议的版本，则路由器会默认发送版本1的消息包。配置RIP路由时，主要涉及路由协议的配置、RIP的路由汇总、RIP计时器以及RIP路由协议认证。

1. 配置RIP路由

步骤1 创建RIP路由进程。
router(config)#**router rip**

步骤2 配置RIP的版本号。
router(config-router)#**version {1 | 2}**

步骤3 定义与RIP路由进程关联的网络。network-number表示直连网段地址。
router(config-router)#**network** *network-number*

第2章 路 由 技 术

步骤4 关闭路由自动汇总。

router(config-router)#**no auto-summary**

步骤5 修改RIP计时器。其中update为RIP的更新定时器，invalid为RIP的失效定时器，flush为RIP的刷新定时器。

router(config-router)#**times basic** *update invalid flush*

步骤6 配置RIP V2的手动汇总。summary-address表示汇总网络地址。netmask表示汇总网络地址段的子网掩码。

router(config-if)#**ip summary-address rip** *summary-address netmask*

2. 配置RIP路由协议认证

步骤1 配置密钥链。key-chain-link表示密钥链。

router(config)#**key chain** *key-chain-link*

步骤2 配置密钥ID。number表示密钥ID。

router(config-keychain)#**key** *number*

步骤3 配置密钥值。value表示密钥值。

router(config-keychain-key)#**key-string** *value*

步骤4 进入相应接口，配置验证方式。

router(config)#**interface** *interface-id*
router(config-if)#**ip rip authenticatin mode md5**

步骤5 启用RIP验证。

router(config-if)#**ip rip authenticatin key-chain** *key-chain-link*

任 务 单

1	RIP V2路由配置
2	RIP路由汇总、定时器、路由协议验证

解决步骤

根据任务单的安排完成任务。

任务1：RIP V2路由配置

任务实施

1. 任务描述及网络拓扑设计

在RA、RB与RC路由器上配置RIP V2动态路由，实现全网互通。绘制拓扑结构图，如图2-3所示。

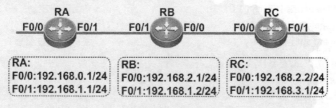

图2-3　RIP V2动态路由

2. 网络设备配置

（1）配置路由器各接口的IP地址

1）在RA路由器上配置IP地址。
RA(config)#interface fastEthernet 0/0
RA(config-if-FastEthernet 0/0)#ip address 192.168.0.1 255.255.255.0
RA(config-if-FastEthernet 0/0)#no shutdown
RA(config-if-FastEthernet 0/0)#exit
RA(config)#interface fastEthernet 0/1
RA(config-if-FastEthernet 0/1)#ip address 192.168.1.1 255.255.255.0
RA(config-if-FastEthernet 0/1)#no shutdown
RA(config-if-FastEthernet 0/1)#exit

2）在RB路由器上配置IP地址。
RB(config)#interface fastEthernet 0/0
RB(config-if-FastEthernet 0/0)#ip address 192.168.2.1 255.255.255.0
RB(config-if-FastEthernet 0/0)#no shutdown
RB(config-if-FastEthernet 0/0)#exit
RB(config)#interface fastEthernet 0/1
RB(config-if-FastEthernet 0/1)#ip address 192.168.1.2 255.255.255.0
RB(config-if-FastEthernet 0/1)#no shutdown
RB(config-if-FastEthernet 0/1)#exit

3）在RC路由器上配置IP地址。
RC(config)#interface fastEthernet 0/0
RC(config-if-FastEthernet 0/0)#ip address 192.168.2.2 255.255.255.0
RC(config-if-FastEthernet 0/0)#no shutdown
RC(config-if-FastEthernet 0/0)#exit
RC(config)#interface fastEthernet 0/1
RC(config-if-FastEthernet 0/1)#ip address 192.168.3.1 255.255.255.0
RC(config-if-FastEthernet 0/1)#no shutdown
RC(config-if-FastEthernet 0/1)#exit

（2）配置RIP路由

1）在RA路由器上配置RIP 路由。
RA(config)#router rip   #启用RIP路由进程
RA(config-router)#version 2   #定义RIP版本号
RA(config-router)#network 192.168.0.0   #宣告路由
RA(config-router)#network 192.168.1.0   #宣告路由
RA(config-router)#no auto-summary   #关闭自动汇总

2）在RB路由器上配置RIP路由。
RB(config)#router rip   #启用RIP路由进程
RB(config-router)#version 2   #定义RIP版本号
RB(config-router)#network 192.168.1.0   #宣告路由
RB(config-router)#network 192.168.2.0   #宣告路由
RB(config-router)#no auto-summary   #关闭自动汇总

3）在RC路由器上配置RIP 路由。
RC(config)#router rip   #启用RIP路由进程
RC(config-router)#version 2   #定义RIP版本号
RC(config-router)#network 192.168.2.0   #宣告路由

第2章 路由技术

```
RC(config-router)#network 192.168.3.0              #宣告路由
RC(config-router)#no auto-summary                  #关闭自动汇总
```

（3）验证配置

1）查看路由表。

```
RA#show ip route                                   #查看路由表
```

```
Codes:  C - connected, S - static, R - RIP, B - BGP
        O - OSPF, IA - OSPF inter area
        N1 - OSPF NSSA external type 1, N2 - OSPF NSSA external type 2
        E1 - OSPF external type 1, E2 - OSPF external type 2
        i - IS-IS, su - IS-IS summary, L1 - IS-IS level-1, L2 - IS-IS level-2
        ia - IS-IS inter area, * - candidate default

Gateway of last resort is no set
C     192.168.0.0/24 is directly connected, FastEthernet 0/0
C     192.168.0.1/32 is local host.
C     192.168.1.0/24 is directly connected, FastEthernet 0/1
C     192.168.1.1/32 is local host.
C     192.168.2.0/24 [120/1] via 192.168.1.2, 00:03:54, FastEthernet 0/1
C     192.168.3.0/24 [120/2] via 192.168.1.2, 00:01:08, FastEthernet 0/1
```

2）连通性测试。

```
RA#ping 192.168.0.1
Sending 5, 100-byte ICMP Echoes to 192.168.0.1, timeout is 2 seconds:
   < press Ctrl+C to break >
!!!!!
Success rate is 100 percent (5/5), round-trip min/avg/max = 1/1/1 ms
RA#ping 192.168.1.1
Sending 5, 100-byte ICMP Echoes to 192.168.1.1, timeout is 2 seconds:
   < press Ctrl+C to break >
!!!!!
Success rate is 100 percent (5/5), round-trip min/avg/max = 1/2/10 ms
RA#ping 192.168.2.1
Sending 5, 100-byte ICMP Echoes to 192.168.2.1, timeout is 2 seconds:
   < press Ctrl+C to break >
!!!!!
Success rate is 100 percent (5/5), round-trip min/avg/max = 1/1/1 ms
RA#ping 192.168.2.2
Sending 5, 100-byte ICMP Echoes to 192.168.2.2, timeout is 2 seconds:
   < press Ctrl+C to break >
!!!!!
Success rate is 100 percent (5/5), round-trip min/avg/max = 1/2/10 ms
RA#ping 192.168.3.1
Sending 5, 100-byte ICMP Echoes to 192.168.3.1, timeout is 2 seconds:
   < press Ctrl+C to break >
!!!!!
Success rate is 100 percent (5/5), round-trip min/avg/max = 1/1/1 ms
```

任务2：配置RIP路由汇总、定时器及路由协议验证

任务实施

1. 任务描述及网络拓扑设计

在RA、RB与RC路由器上配置RIP V2动态路由，实现全网互通；为了减少路由器查找路由表的时间，希望减小路由表的规模，在RA与RC路由器上配置RIP手工汇总；出于安全性的考虑，希望路由信息交换都是在可信任的路由器之间进行，可以通过在路由器上配置路由协议认证实现。绘制拓扑结构图，如图2-4所示。

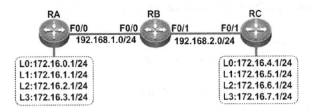

图2-4　RIP路由汇总

2. 网络设备配置

（1）配置路由器各接口的IP地址

1）在RA路由器上配置IP地址。
RA(config)#interface fastEthernet 0/0
RA(config-if-FastEthernet 0/0)#ip address 192.168.1.1 255.255.255.0
RA(config-if-FastEthernet 0/0)#no shutdown
RA(config-if-FastEthernet 0/0)#exit
RA(config)#interface loopback 0
RA(config-if-Loopback 0)#ip address 172.16.0.1 255.255.255.0
RA(config-if-Loopback 0)#exit
RA(config)#interface loopback 1
RA(config-if-Loopback 1)#ip address 172.16.1.1 255.255.255.0
RA(config-if-Loopback 1)#exit
RA(config)#interface loopback 2
RA(config-if-Loopback 2)#ip address 172.16.2.1 255.255.255.0
RA(config-if-Loopback 2)#exit
RA(config)#interface loopback 3
RA(config-if-Loopback 3)#ip address 172.16.3.1 255.255.255.0
RA(config-if-Loopback 3)#exit

2）在RB路由器上配置IP地址。
RB(config)#interface fastEthernet 0/0
RB(config-if-FastEthernet 0/0)#ip address 192.168.1.2 255.255.255.0
RB(config-if-FastEthernet 0/0)#no shutdown
RB(config-if-FastEthernet 0/0)#exit
RB(config)#interface fastEthernet 0/1
RB(config-if-FastEthernet 0/1)#ip address 192.168.2.1 255.255.255.0
RB(config-if-FastEthernet 0/1)#no shutdown
RB(config-if-FastEthernet 0/1)#exit

3）在RC路由器上配置IP地址。
RC(config)#interface fastEthernet 0/1
RC(config-if-FastEthernet 0/1)#ip address 192.168.2.2 255.255.255.0
RC(config-if-FastEthernet 0/1)#no shutdown
RC(config-if-FastEthernet 0/1)#exit
RC(config)#interface loopback 0
RC(config-if-Loopback 0)#ip address 172.16.4.1 255.255.255.0
RC(config-if-Loopback 0)#exit
RC(config)#interface loopback 1
RC(config-if-Loopback 1)#ip address 172.16.5.1 255.255.255.0
RC(config-if-Loopback 1)#exit
RC(config)#interface loopback 2
RC(config-if-Loopback 2)#ip address 172.16.6.1 255.255.255.0
RC(config-if-Loopback 2)#exit
RC(config)#interface loopback 3
RC(config-if-Loopback 3)#ip address 172.16.7.1 255.255.255.0
RC(config-if-Loopback 3)#

（2）配置RIP版本

1）在RA路由器上配置RIP路由。
RA(config)#router rip
RA(config-router)#version 2
RA(config-router)#network 172.16.0.0
RA(config-router)#network 192.168.1.0
RA(config-router)#no auto-summary

2）在RB路由器上配置RIP路由。
RB(config)#router rip
RB(config-router)#version 2
RB(config-router)#network 192.168.1.0
RB(config-router)#network 192.168.2.0
RB(config-router)#no auto-summary

3）在RC路由器上配置RIP路由。
RC(config)#router rip
RC(config-router)#version 2
RC(config-router)#network 192.168.2.0
RC(config-router)#network 172.16.0.0
RC(config-router)#no auto-summary

（3）验证测试
RB#show ip route

Codes: C - connected, S - static, R - RIP, B - BGP
         O - OSPF, IA - OSPF inter area
         N1 - OSPF NSSA external type 1, N2 - OSPF NSSA external type 2
         E1 - OSPF external type 1, E2 - OSPF external type 2
         i - IS-IS, su - IS-IS summary, L1 - IS-IS level-1, L2 - IS-IS level-2
         ia - IS-IS inter area, * - candidate default

Gateway of last resort is no set
R   172.16.0.0/24 [120/1] via 192.168.1.1, 00:10:05, FastEthernet 0/0

R   172.16.1.0/24 [120/1] via 192.168.1.1, 00:10:05, FastEthernet 0/0
R   172.16.2.0/24 [120/1] via 192.168.1.1, 00:10:05, FastEthernet 0/0
R   172.16.3.0/24 [120/1] via 192.168.1.1, 00:10:05, FastEthernet 0/0
R   172.16.4.0/24 [120/1] via 192.168.2.2, 00:06:24, FastEthernet 0/1
R   172.16.4.0/22 [120/1] via 192.168.2.2, 00:01:24, FastEthernet 0/1
R   172.16.5.0/24 [120/1] via 192.168.2.2, 00:06:24, FastEthernet 0/1
R   172.16.6.0/24 [120/1] via 192.168.2.2, 00:06:24, FastEthernet 0/1
R   172.16.7.0/24 [120/1] via 192.168.2.2, 00:06:24, FastEthernet 0/1
C   192.168.1.0/24 is directly connected, FastEthernet 0/0
C   192.168.1.2/32 is local host.
C   192.168.2.0/24 is directly connected, FastEthernet 0/1
C   192.168.2.1/32 is local host.

（4）配置RIP手动汇总

1）在RA路由器上配置手动汇总。
RA(config)#interface fastEthernet 0/0
RA(config-if-FastEthernet 0/0)#ip summary-address rip 172.16.0.0 255.255.252.0

2）在RC路由器上配置手动汇总。
RC(config)#interface fastEthernet 0/1
RC(config-if-FastEthernet 0/1)#ip summary-address rip 172.16.4.0 255.255.252.0

（5）验证测试
RB#show ip route
Codes: C - connected, S - static, R - RIP, B - BGP
        O - OSPF, IA - OSPF inter area
        N1 - OSPF NSSA external type 1, N2 - OSPF NSSA external type 2
        E1 - OSPF external type 1, E2 - OSPF external type 2
        i - IS-IS, su - IS-IS summary, L1 - IS-IS level-1, L2 - IS-IS level-2
        ia - IS-IS inter area, * - candidate default

Gateway of last resort is no set
R   172.16.0.0/22 [120/1] via 192.168.1.1, 00:03:29, FastEthernet 0/0
R   172.16.4.0/22 [120/1] via 192.168.2.2, 00:09:10, FastEthernet 0/1
C   192.168.1.0/24 is directly connected, FastEthernet 0/0
C   192.168.1.2/32 is local host.
C   192.168.2.0/24 is directly connected, FastEthernet 0/1
C   192.168.2.1/32 is local host.

（6）配置定时器

1）在RA路由器上配置定时器。
RA(config)#router rip
RA(config-router)#timers basic 20 120 80
        #修改RIP的更新定时器、失效定时器和刷新定时器的默认值

2）在RB路由器上配置定时器。
RB(config)#router rip
RB(config-router)#timers basic 20 120 80   #配置RIP定时器

3）在RC路由器上配置定时器。
RC(config)#router rip
RC(config-router)#timers basic 20 120 80   #配置RIP定时器
RC(config-router)#

第2章 路由技术

（7）配置RIP验证

1）在RA路由器上配置RIP验证。

```
RA(config)#key chain ruijie                              #配置密钥链
RA(config-keychain)#key 1                                #配置密钥ID
RA(config-keychain-key)#key-string 12345                 #配置密钥值
RA(config-keychain-key)#exit
RA(config-keychain)#exit
RA(config)#interface fastEthernet 0/0
RA(config-if)#ip rip authentication mode md5            #配置验证方式为MD5
RA(config-if)#ip rip authentication key-chain ruijie    #配置接口启用RIP验证
```

2）在RB路由器上配置RIP验证。

```
RB(config)#key chain ruijie                              #配置密钥链
RB(config-keychain)#key 1                                #配置密钥ID
RB(config-keychain-key)#key-string 12345                 #配置密钥值
RB(config-keychain-key)#exit
RB(config-keychain)#exit
RB(config)#interface fastEthernet 0/0
RB(config-if)#ip rip authentication mode md5            #配置验证方式为MD5
RB(config-if)#ip rip authentication key-chain ruijie    #配置接口启用RIP验证
RB(config-if)#exit
RB(config)#interface fastEthernet 0/1
RB(config-if)#ip rip authentication mode md5            #配置验证方式为MD5
RB(config-if)#ip rip authentication key-chain ruijie    #在接口上应用密钥链
```

3）在RC路由器上配置RIP验证。

```
RC(config)#key chain ruijie                              #配置密钥链
RC(config-keychain)#key 1                                #配置密钥ID
RC(config-keychain-key)#key-string 12345                 #配置密钥值
RC(config-keychain-key)#exit
RC(config-keychain)#exit
RC(config)#interface fastEthernet 0/1
RC(config-if)#ip rip authentication mode md5            #配置验证方式为MD5
RC(config-if)#ip rip authentication key-chain ruijie    #在接口上应用密钥链
```

小结

通过RIP路由协议的学习，主要掌握RIP V2路由、RIP路由汇总和RIP路由协议认证等配置与管理工作。

2.3 OSPF路由

问题描述

主校区采用双核心双出口架构，对网络出口和网络核心实现高可用性和高可靠性，所以在路由规划时,主校区的网络选路采用OSPF路由协议。OSPF路由协议适合于大中型网络规模,是基于层次型网络架构的路由协议,其收敛速度快,而且又是一个无环路的路由协议。为保障路由更新的安全性,OSPF路由协议需要进行基于接口的MD5验证。

问题分析

配置OSPF路由时,主要涉及OSPF单区域配置、OSPF多区域配置、OSPF路由汇总,以及OSPF路由协议认证。

1. 配置OSPF

步骤1 创建OSPF路由进程。

router(config)#**router ospf** *process-id*

步骤2 定义接口所属区域。address为网络地址,inverse-mask为子网掩码反码,area-id是区域编号。

router(config-router)#**network** *address inverse-mask* **area** *area-id*

步骤3 指定接口的花费。

router(config-if)#**ip ospf cost** *cost*

步骤4 配置Router ID。

router(config-router)#**router-id** *ip-address*

2. 配置OSPF路由汇总

步骤1 配置OSPF内部路由汇总。area-id是区域编号。

router(config-router)#**area** *area-id* **range** *network mask*

步骤2 配置OSPF外部路由汇总。

router(config-router)#**summary-address** *network mask*

3. 配置OSPF路由协议认证

步骤1 设置OSPF明文认证密钥。key是密钥。

router(config-if)#**ip ospf authentication-key** *key*

步骤2 设置OSPF MD5认证密钥。key-id为密钥ID,key为密钥。

router(config-if)#**ip ospf message-digest-key** *key-id* **md5** *key*

任 务 单

1	OSPF单区域路由配置
2	OSPF多区域路由配置
3	OSPF区域汇总
4	配置OSP路由协议认证

解决步骤

根据任务单的安排完成任务。

第2章 路由技术

任务1：OSPF单区域路由配置

任务实施

1. 任务描述及网络拓扑设计

在RA、RB与RC路由器上配置OSPF单区域路由，实现全网互通。绘制拓扑结构图，如图2-5所示。

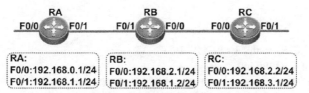

图2-5　OSPF单区域路由

2. 网络设备配置

（1）配置路由器各接口的IP地址

1）在RA路由器上配置IP地址。
RA(config)#interface fastEthernet 0/0
RA(config-if-FastEthernet 0/0)#ip address 192.168.0.1 255.255.255.0
RA(config-if-FastEthernet 0/0)#no shutdown
RA(config-if-FastEthernet 0/0)#exit
RA(config)#interface fastEthernet 0/1
RA(config-if-FastEthernet 0/1)#ip address 192.168.1.1 255.255.255.0
RA(config-if-FastEthernet 0/1)#exit

2）在RB路由器上配置IP地址。
RB(config)#interface fastEthernet 0/1
RB(config-if-FastEthernet 0/1)#ip address 192.168.1.2 255.255.255.0
RB(config-if-FastEthernet 0/1)#no shutdown
RB(config-if-FastEthernet 0/1)#exit
RB(config)#interface fastEthernet 0/0
RB(config-if-FastEthernet 0/0)#ip address 192.168.2.1 255.255.255.0
RB(config-if-FastEthernet 0/0)#no shutdown

3）在RC路由器上配置IP地址。
RC(config)#interface fastEthernet 0/0
RC(config-if-FastEthernet 0/0)#ip address 192.168.2.2 255.255.255.0
RC(config-if-FastEthernet 0/0)#no shutdown
RC(config-if-FastEthernet 0/0)#exit
RC(config)#interface fastEthernet 0/1
RC(config-if-FastEthernet 0/1)#ip address 192.168.3.1 255.255.255.0
RC(config-if-FastEthernet 0/1)#no shutdown

（2）配置OSPF

1）在RA路由器上配置OSPF路由。
RA(config)#router ospf 10   #启用OSPF路由进程
RA(config-router)#network 192.168.0.0 0.0.0.255 area 0   #宣告直连路由
RA(config-router)#network 192.168.1.0 0.0.0.255 area 0   #宣告直连路由

2）在RB路由器上配置OSPF路由。
```
RB(config)#router ospf 10                                      #启用OSPF路由进程
RB(config-router)#network 192.168.1.0 0.0.0.255 area 0         #宣告直连路由
RB(config-router)#network 192.168.2.0 0.0.0.255 area 0         #宣告直连路由
```
3）在RC路由器上配置OSPF路由。
```
RC(config)#router ospf 10                                      #启用OSPF路由进程
RC(config-router)#network 192.168.2.0 0.0.0.255 area 0         #宣告直连路由
RC(config-router)#network 192.168.3.0 0.0.0.255 area 0         #宣告直连路由
```

（3）验证测试

```
RB#show ip route

Codes:   C - connected, S - static, R - RIP, B - BGP
         O - OSPF, IA - OSPF inter area
         N1 - OSPF NSSA external type 1, N2 - OSPF NSSA external type 2
         E1 - OSPF external type 1, E2 - OSPF external type 2
         i - IS-IS, su - IS-IS summary, L1 - IS-IS level-1, L2 - IS-IS level-2
         ia - IS-IS inter area, * - candidate default

Gateway of last resort is no set
O     192.168.0.0/24 [110/2] via 192.168.1.1, 00:04:26, FastEthernet 0/1
C     192.168.1.0/24 is directly connected, FastEthernet 0/1
C     192.168.1.2/32 is local host.
C     192.168.2.0/24 is directly connected, FastEthernet 0/0
C     192.168.2.1/32 is local host.
O     192.168.3.0/24 [110/2] via 192.168.2.2, 00:02:01, FastEthernet 0/0
```

从RB的路由表可以看出，RA通过OSPF学习到了192.168.1.0/24和192.168.3.0/24网段的路由信息。

```
RB#show ip ospf neighbor

OSPF process 1, 2 Neighbors, 2 is Full:
Neighbor ID    Pri   State          BFD State    Dead Time    Address
   Interface
192.168.1.1    1     Full/DR        -            00:00:39     192.168.1.1
   FastEthernet 0/1
192.168.3.1    1     Full/BDR       -            00:00:39     192.168.2.2
   FastEthernet 0/0
```

从显示信息可以看出，RB与RA和RC建立了FULL的邻接关系。

任务2：OSPF多区域路由配置

任务实施

1. 任务描述及网络拓扑设计

在RA、RB与RC路由器上配置OSPF多区域路由，实现全网互通。其中RA路由器的L0和L1接口在Area1区域，F0/0接口在Area0区域；RB路由器的L0、L1和F0/0接口在Area0区域，F0/1接口在Area2区域；RC路由器的L0、L1和F0/1接口在Area2区域。绘制拓扑结构图，如图2-6所示。

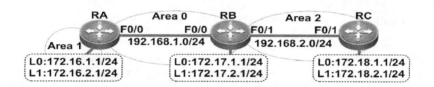

图2-6 OSPF多区域路由

2. 网络设备配置

（1）配置路由器各接口的IP地址

1）在RA路由器上配置IP地址。

RA(config)#interface fastEthernet 0/0
RA(config-if-FastEthernet 0/0)#ip address 192.168.1.1 255.255.255.0
RA(config-if-FastEthernet 0/0)#no shutdown
RA(config-if-FastEthernet 0/0)#exit
RA(config)#interface loopback 0
RA(config-if-Loopback 0)#ip address 172.16.1.1 255.255.255.0
RA(config-if-Loopback 0)#exit
RA(config)#interface loopback 1
RA(config-if-Loopback 1)#ip address 172.16.2.1 255.255.255.0

2）在RB路由器上配置IP地址。

RB(config)#interface fastEthernet 0/0
RB(config-if-FastEthernet 0/0)#ip address 192.168.1.2 255.255.255.0
RB(config-if-FastEthernet 0/0)#no shutdown
RB(config-if-FastEthernet 0/0)#exit
RB(config)#interface fastEthernet 0/1
RB(config-if-FastEthernet 0/1)#ip address 192.168.2.1 255.255.255.0
RB(config-if-FastEthernet 0/1)#no shutdown
RB(config-if-FastEthernet 0/1)#exit
RB(config)#interface loopback 0
RB(config-if-Loopback 0)#ip address 172.17.1.1 255.255.255.0
RB(config-if-Loopback 0)#exit
RB(config)#interface loopback 1
RB(config-if-Loopback 1)#ip address 172.17.2.1 255.255.255.0
RB(config-if-Loopback 1)#exit

3）在RC路由器上配置IP地址。

RC(config)#interface fastEthernet 0/1
RC(config-if-FastEthernet 0/1)#ip address 192.168.2.2 255.255.255.0
RC(config-if-FastEthernet 0/1)#no shutdown
RC(config-if-FastEthernet 0/1)#exit
RC(config)#interface loopback 0
RC(config-if-Loopback 0)#ip address 172.18.1.1 255.255.255.0
RC(config-if-Loopback 0)#exit
RC(config)#interface loopback 1

RC(config-if-Loopback 1)#ip address 172.18.2.1 255.255.255.0
RC(config-if-Loopback 1)#exit

（2）配置OSPF

1）在RA路由器上配置OSPF路由。

RA(config)#router ospf   #启用OSPF路由进程
RA(config-router)#network 192.168.1.0 0.0.0.255 area 0   #宣告直连路由
RA(config-router)#network 172.16.1.0 0.0.0.255 area 1   #宣告直连路由
RA(config-router)#network 172.16.2.0 0.0.0.255 area 1   #宣告直连路由

2）在RB路由器上配置OSPF路由。

RB(config)#router ospf   #启用OSPF路由进程
RB(config-router)#network 192.168.1.0 0.0.0.255 area 0   #宣告直连路由
RB(config-router)#network 172.17.1.0 0.0.0.255 area 0   #宣告直连路由
RB(config-router)#network 172.17.2.0 0.0.0.255 area 0   #宣告直连路由
RB(config-router)#network 192.168.2.0 0.0.0.255 area 2   #宣告直连路由

3）在RC路由器上配置OSPF路由。

RC(config)#router ospf   #启用OSPF路由进程
RC(config-router)#network 192.168.2.0 0.0.0.255 area 2   #宣告直连路由
RC(config-router)#network 172.18.1.0 0.0.0.255 area 2   #宣告直连路由
RC(config-router)#network 172.18.2.0 0.0.0.255 area 2   #宣告直连路由

（3）验证测试

1）在RA路由器上查看路由表。

RA#show ip route

Codes: C - connected, S - static, R - RIP, B - BGP
          O - OSPF, IA - OSPF inter area
          N1 - OSPF NSSA external type 1, N2 - OSPF NSSA external type 2
          E1 - OSPF external type 1, E2 - OSPF external type 2
          i - IS-IS, su - IS-IS summary, L1 - IS-IS level-1, L2 - IS-IS level-2
          ia - IS-IS inter area, * - candidate default

Gateway of last resort is no set
C   172.16.1.0/24 is directly connected, Loopback 0
C   172.16.1.1/32 is local host.
C   172.16.2.0/24 is directly connected, Loopback 1
C   172.16.2.1/32 is local host.
O   172.17.1.1/32 [110/1] via 192.168.1.2, 00:11:34, FastEthernet 0/0
O   172.17.2.1/32 [110/1] via 192.168.1.2, 00:11:24, FastEthernet 0/0
O IA   172.18.1.1/32 [110/2] via 192.168.1.2, 00:01:34, FastEthernet 0/0
O IA   172.18.2.1/32 [110/2] via 192.168.1.2, 00:01:34, FastEthernet 0/0
C   192.168.1.0/24 is directly connected, FastEthernet 0/0
C   192.168.1.1/32 is local host.
O IA   192.168.2.0/24 [110/2] via 192.168.1.2, 00:01:53, FastEthernet 0/0

从RA的路由表可以看出，RA通过OSPF区域内路由学习到了172.17.1.1/32和172.17.2.1/32网段的路由信息，通过OSPF区域间路由学习到了172.18.1.1/32、172.18.2.1/32和192.168.2.0/24的路由信息。

2）在RB路由器上查看路由表。

```
RB#show ip route

Codes:  C - connected, S - static, R - RIP, B - BGP
        O - OSPF, IA - OSPF inter area
        N1 - OSPF NSSA external type 1, N2 - OSPF NSSA external type 2
        E1 - OSPF external type 1, E2 - OSPF external type 2
        i - IS-IS, su - IS-IS summary, L1 - IS-IS level-1, L2 - IS-IS level-2
        ia - IS-IS inter area, * - candidate default

Gateway of last resort is no set
O IA 172.16.1.1/32 [110/1] via 192.168.1.1, 00:10:21, FastEthernet 0/0
O IA 172.16.2.1/32 [110/1] via 192.168.1.1, 00:10:21, FastEthernet 0/0
C    172.17.1.0/24 is directly connected, Loopback 0
C    172.17.1.1/32 is local host.
C    172.17.2.0/24 is directly connected, Loopback 1
C    172.17.2.1/32 is local host.
O    172.18.1.1/32 [110/1] via 192.168.2.2, 00:00:20, FastEthernet 0/1
O    172.18.2.1/32 [110/1] via 192.168.2.2, 00:00:20, FastEthernet 0/1
C    192.168.1.0/24 is directly connected, FastEthernet 0/0
C    192.168.1.2/32 is local host.
C    192.168.2.0/24 is directly connected, FastEthernet 0/1
C    192.168.2.1/32 is local host.
```

从RB的路由表可以看出，RB通过OSPF区域内路由学习到了172.18.1.1/32和172.18.2.1/32网段的路由信息，通过OSPF区域间路由学习到了172.16.1.1/32和172.16.2.1/32的路由信息。

3）在RC路由器上查看路由表。

```
RC#show ip route

Codes:  C - connected, S - static, R - RIP, B - BGP
        O - OSPF, IA - OSPF inter area
        N1 - OSPF NSSA external type 1, N2 - OSPF NSSA external type 2
        E1 - OSPF external type 1, E2 - OSPF external type 2
        i - IS-IS, su - IS-IS summary, L1 - IS-IS level-1, L2 - IS-IS level-2
        ia - IS-IS inter area, * - candidate default

Gateway of last resort is no set
O IA 172.16.1.1/32 [110/2] via 192.168.2.1, 00:02:18, FastEthernet 0/1
O IA 172.16.2.1/32 [110/2] via 192.168.2.1, 00:02:18, FastEthernet 0/1
O IA 172.17.1.1/32 [110/1] via 192.168.2.1, 00:02:18, FastEthernet 0/1
O IA 172.17.2.1/32 [110/1] via 192.168.2.1, 00:02:18, FastEthernet 0/1
C    172.18.1.0/24 is directly connected, Loopback 0
C    172.18.1.1/32 is local host.
C    172.18.2.0/24 is directly connected, Loopback 1
C    172.18.2.1/32 is local host.
O IA 192.168.1.0/24 [110/2] via 192.168.2.1, 00:02:18, FastEthernet 0/1
C    192.168.2.0/24 is directly connected, FastEthernet 0/1
C    192.168.2.2/32 is local host.
```

从RC的路由表可以看出，RC通过OSPF区域间路由学习到了172.16.1.1/32、

172.16.2.1/32、172.17.1.1/32、172.17.2.1/32和192.168.1.0/24的路由信息。

任务3：配置OSPF路由汇总

任务实施

1. 任务描述及网络拓扑设计

在RA、RB与RC路由器上配置OSPF多区域路由，实现全网互通。其中RA路由器的L0、L1、L2和L3接口在Area1区域，F0/0接口在Area0区域；RB路由器的F0/0接口在Area0区域，F0/1接口在Area2区域；RC路由器的F0/1接口在Area2区域，L0、L1、L2和L3接口为外部路由。为了减少路由表条目，需要在RA路由器上配置路由汇总，在RC路由器上配置区域外路由汇总。绘制拓扑结构图，如图2-7所示。

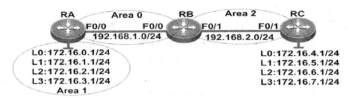

图2-7　OSPF路由汇总

2. 网络设备配置

（1）配置路由器各接口的IP地址

1）在RA路由器上配置IP地址。
```
RA(config)#interface fastEthernet 0/0
RA(config-if-FastEthernet 0/0)#ip address 192.168.1.1 255.255.255.0
RA(config-if-FastEthernet 0/0)#no shutdown
RA(config-if-FastEthernet 0/0)#exit
RA(config)#interface loopback 0
RA(config-if-Loopback 0)#ip address 172.16.0.1 255.255.255.0
RA(config-if-Loopback 0)#exit
RA(config)#interface loopback 1
RA(config-if-Loopback 1)#ip address 172.16.1.1 255.255.255.0
RA(config-if-Loopback 1)#exit
RA(config)#interface loopback 2
RA(config-if-Loopback 2)#ip address 172.16.2.1 255.255.255.0
RA(config-if-Loopback 2)#exit
RA(config)#interface loopback 3
RA(config-if-Loopback 3)#ip address 172.16.3.1 255.255.255.0
RA(config-if-Loopback 3)#exit
```

2）在RB路由器上配置IP地址。
```
RB(config)#interface fastEthernet 0/0
RB(config-if-FastEthernet 0/0)#ip address 192.168.1.2 255.255.255.0
RB(config-if-FastEthernet 0/0)#no shutdown
RB(config-if-FastEthernet 0/0)#exit
RB(config)#interface fastEthernet 0/1
```

```
RB(config-if-FastEthernet 0/1)#ip address 192.168.2.1 255.255.255.0
RB(config-if-FastEthernet 0/1)#no shutdown
RB(config-if-FastEthernet 0/1)#exit
```

3）在RC路由器上配置IP地址。

```
RC(config)#interface fastEthernet 0/1
RC(config-if-FastEthernet 0/1)#ip address 192.168.2.2 255.255.255.0
RC(config-if-FastEthernet 0/1)#no shutdown
RC(config-if-FastEthernet 0/1)#exit
RC(config)#interface loopback 0
RC(config-if-Loopback 0)#ip address 172.16.4.1 255.255.255.0
RC(config-if-Loopback 0)#exit
RC(config)#interface loopback 1
RC(config-if-Loopback 1)#ip address 172.16.5.1 255.255.255.0
RC(config-if-Loopback 1)#exit
RC(config)#interface loopback 2
RC(config-if-Loopback 2)#ip address 172.16.6.1 255.255.255.0
RC(config-if-Loopback 2)#exit
RC(config)#interface loopback 3
RC(config-if-Loopback 3)#ip address 172.16.7.1 255.255.255.0
```

（2）配置OSPF

1）在RA路由器上配置OSPF路由。

```
RA(config)#router ospf
RA(config-router)#network 192.168.1.0 0.0.0.255 area 0
RA(config-router)#network 172.16.0.0 0.0.0.255 area 1
RA(config-router)#network 172.16.1.0 0.0.0.255 area 1
RA(config-router)#network 172.16.2.0 0.0.0.255 area 1
RA(config-router)#network 172.16.3.0 0.0.0.255 area 1
```

2）在RB路由器上配置OSPF路由。

```
RB(config)#router ospf
RB(config-router)#network 192.168.1.0 0.0.0.255 area 0
RB(config-router)#network 192.168.2.0 0.0.0.255 area 2
```

3）在RC路由器上配置OSPF路由。

```
RC(config)#router ospf
RC(config-router)#network 192.168.2.0 0.0.0.255 area 2
```

（3）验证测试

1）在RC路由器上查看路由表。

```
RC#show ip route
Codes:   C - connected, S - static, R - RIP, B - BGP
         O - OSPF, IA - OSPF inter area
         N1 - OSPF NSSA external type 1, N2 - OSPF NSSA external type 2
         E1 - OSPF external type 1, E2 - OSPF external type 2
         i - IS-IS, su - IS-IS summary, L1 - IS-IS level-1, L2 - IS-IS level-2
         ia - IS-IS inter area, * - candidate default

Gateway of last resort is no set
O IA 172.16.0.1/32 [110/2] via 192.168.2.1, 00:01:04, FastEthernet 0/1
O IA 172.16.1.1/32 [110/2] via 192.168.2.1, 00:01:04, FastEthernet 0/1
```

O IA 172.16.2.1/32 [110/2] via 192.168.2.1, 00:01:04, FastEthernet 0/1
O IA 172.16.3.1/32 [110/2] via 192.168.2.1, 00:01:04, FastEthernet 0/1
C   172.16.4.0/24 is directly connected, Loopback 0
C   172.16.4.1/32 is local host.
C   172.16.5.0/24 is directly connected, Loopback 1
C   172.16.5.1/32 is local host.
C   172.16.6.0/24 is directly connected, Loopback 2
C   172.16.6.1/32 is local host.
C   172.16.7.0/24 is directly connected, Loopback 3
C   172.16.7.1/32 is local host.
O IA 192.168.1.0/24 [110/2] via 192.168.2.1, 00:01:04, FastEthernet 0/1
C   192.168.2.0/24 is directly connected, FastEthernet 0/1
C   192.168.2.2/32 is local host.

从RC的路由表可以看出，RC学习到了RA的4个Loopback接口的路由信息。

2）在RA路由器上查看路由表。

RA#show ip route

Codes: C - connected, S - static, R - RIP, B - BGP
        O - OSPF, IA - OSPF inter area
        N1 - OSPF NSSA external type 1, N2 - OSPF NSSA external type 2
        E1 - OSPF external type 1, E2 - OSPF external type 2
        i - IS-IS, su - IS-IS summary, L1 - IS-IS level-1, L2 - IS-IS level-2
        ia - IS-IS inter area, * - candidate default

Gateway of last resort is no set

C   172.16.0.0/24 is directly connected, Loopback 0
C   172.16.0.1/32 is local host.
C   172.16.1.0/24 is directly connected, Loopback 1
C   172.16.1.1/32 is local host.
C   172.16.2.0/24 is directly connected, Loopback 2
C   172.16.2.1/32 is local host.
C   172.16.3.0/24 is directly connected, Loopback 3
C   172.16.3.1/32 is local host.
C   192.168.1.0/24 is directly connected, FastEthernet 0/0
C   192.168.1.1/32 is local host.
O IA 192.168.2.0/24 [110/2] via 192.168.1.2, 00:06:58, FastEthernet 0/0

从RA的路由表可以看出，RA没有学习到RC的Loopback接口的路由信息。

（4）直连路由重分发

RC(config)#router ospf
RC(config-router)#redistribute connected subnets metric 3   #配置直连路由重分发

（5）验证测试

RA#show ip route

Codes: C - connected, S - static, R - RIP, B - BGP
        O - OSPF, IA - OSPF inter area
        N1 - OSPF NSSA external type 1, N2 - OSPF NSSA external type 2
        E1 - OSPF external type 1, E2 - OSPF external type 2
        i - IS-IS, su - IS-IS summary, L1 - IS-IS level-1, L2 - IS-IS level-2
        ia - IS-IS inter area, * - candidate default

```
Gateway of last resort is no set
C       172.16.0.0/24 is directly connected, Loopback 0
C       172.16.0.1/32 is local host.
C       172.16.1.0/24 is directly connected, Loopback 1
C       172.16.1.1/32 is local host.
C       172.16.2.0/24 is directly connected, Loopback 2
C       172.16.2.1/32 is local host.
C       172.16.3.0/24 is directly connected, Loopback 3
C       172.16.3.1/32 is local host.
O E2 172.16.4.0/24 [110/3] via 192.168.1.2, 00:00:44, FastEthernet 0/0
O E2 172.16.5.0/24 [110/3] via 192.168.1.2, 00:00:44, FastEthernet 0/0
O E2 172.16.6.0/24 [110/3] via 192.168.1.2, 00:00:44, FastEthernet 0/0
O E2 172.16.7.0/24 [110/3] via 192.168.1.2, 00:00:44, FastEthernet 0/0
C       192.168.1.0/24 is directly connected, FastEthernet 0/0
C       192.168.1.1/32 is local host.
O IA 192.168.2.0/24 [110/2] via 192.168.1.2, 00:09:36, FastEthernet 0/0
```

从RA的路由表可以看出，RA学习到了RC的Loopback接口的外部路由信息。

（6）在RA路由器上配置区域间路由汇总

```
RA(config)#router ospf
RA(config-router)#area 1 range 172.16.0.0 255.255.252.0      #配置区域间路由汇总
```

（7）在RC路由器上配置OSPF外部路由汇总

```
RC(config-router)#summary-address 172.16.4.0 255.255.252.0   #配置外部路由汇总
```

（8）验证测试

```
RB#show ip route

Codes:  C - connected, S - static, R - RIP, B - BGP
        O - OSPF, IA - OSPF inter area
        N1 - OSPF NSSA external type 1, N2 - OSPF NSSA external type 2
        E1 - OSPF external type 1, E2 - OSPF external type 2
        i - IS-IS, su - IS-IS summary, L1 - IS-IS level-1, L2 - IS-IS level-2
        ia - IS-IS inter area, * - candidate default

Gateway of last resort is no set
O IA 172.16.0.0/22 [110/1] via 192.168.1.1, 00:03:47, FastEthernet 0/0
O E2 172.16.4.0/22 [110/3] via 192.168.2.2, 00:01:29, FastEthernet 0/1
C       192.168.1.0/24 is directly connected, FastEthernet 0/0
C       192.168.1.2/32 is local host.
C       192.168.2.0/24 is directly connected, FastEthernet 0/1
C       192.168.2.1/32 is local host.
```

从RB的路由表可以看到，RB学习到了RA的Loopback接口的汇总路由，学习到了RC的Loopback接口的外部汇总路由。

任务4：配置OSPF路由协议认证

任务实施

1. 任务描述及网络拓扑设计

在RA、RB与RC路由器上配置OSPF多区域路由，实现全网互通。为了确保网络中路由信息的交换安全性，需要控制网络中的路由更新只在可信任的路由器之间学习。在RA与RB之间使用明文验证，在RB与RC之间使用MD5密文验证。绘制拓扑结构图，如图2-8所示。

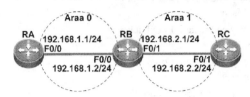

图2-8　OSPF路由协议认证

2. 网络设备配置

（1）配置路由器各接口的IP地址

1）在RA路由器上配置IP地址。

RA(config)#interface fastEthernet 0/0
RA(config-if-FastEthernet 0/0)#ip address 192.168.1.1 255.255.255.0
RA(config-if-FastEthernet 0/0)#no shutdown
RA(config-if-FastEthernet 0/0)#exit

2）在RB路由器上配置IP地址。

RB(config)#interface fastEthernet 0/0
RB(config-if-FastEthernet 0/0)#ip address 192.168.1.2 255.255.255.0
RB(config-if-FastEthernet 0/0)#no shutdown
RB(config-if-FastEthernet 0/0)#exit
RB(config)#interface fastEthernet 0/1
RB(config-if-FastEthernet 0/1)#ip address 192.168.2.1 255.255.255.0
RB(config-if-FastEthernet 0/1)#no shutdown
RB(config-if-FastEthernet 0/1)#exit

3）在RC路由器上配置IP地址。

RC(config)#interface fastEthernet 0/1
RC(config-if-FastEthernet 0/1)#ip address 192.168.2.2 255.255.255.0
RC(config-if-FastEthernet 0/1)#no shutdown

（2）配置OSPF

1）在RA路由器上配置OSPF路由。

RA(config)#router ospf
RA(config-router)#network 192.168.1.0 0.0.0.255 area 0
RA(config-router)#exit

2）在RB路由器上配置OSPF路由。

RB(config)#router ospf
RB(config-router)#network 192.168.1.0 0.0.0.255 area 0
RB(config-router)#network 192.168.2.0 0.0.0.255 area 1

3）在RC路由器上配置OSPF路由。

RC(config)#router ospf
RC(config-router)#network 192.168.2.0 0.0.0.255 area 1

（3）配置OSPF验证

1）在RA路由器的F0/0接口上配置明文验证。

RA(config)#interface fastEthernet 0/0
RA(config-if-FastEthernet 0/0)#ip ospf authentication　　　　　#启用接口的明文验证

RA(config-if-FastEthernet 0/0)#ip ospf authentication-key 123   #配置明文验证的密码

2）在RB路由器的F0/0接口上配置明文验证，F0/1接口上配置密文验证。
RB(config)#interface fastEthernet 0/0
RB(config-if-FastEthernet 0/0)#ip ospf authentication   #启用接口的明文验证
RB(config-if-FastEthernet 0/0)#ip ospf authentication-key 123   #配置明文验证的密码
RB(config-if-FastEthernet 0/0)#exit
RB(config)#interface fastEthernet 0/1
RB(config-if-FastEthernet 0/1)#ip ospf authentication message-digest
        #启用接口的MD5验证
RB(config-if-FastEthernet 0/1)#ip ospf message-digest-key 1 md5 aaa
        #配置MD5验证的密钥ID和密钥

3）在RC路由器的F0/1接口上配置密文验证。
RC(config)#interface fastEthernet 0/1
RC(config-if-FastEthernet 0/1)#ip ospf authentication message-digest
        #启用接口的MD5验证
RC(config-if-FastEthernet 0/1)#ip ospf message-digest-key 1 md5 aaa
        #配置MD5验证的密钥ID和密钥

（4）验证测试
RB#show ip ospf neighbor

OSPF process 1, 2 Neighbors, 2 is Full:
Neighbor ID   Pri   State   BFD State Dead Time Address
  Interface
192.168.1.1   1   Full/BDR   -   00:00:30   192.168.1.1
  FastEthernet 0/0
192.168.2.2   1   Full/DR   -   00:00:30   192.168.2.2
  FastEthernet 0/1

通过以上邻居状态信息可以看到，RB与RA和RC成功地建立了FULL的邻接关系。

小结

通过OSPF路由协议的学习，主要掌握OSPF单区域路由、OSPF多区域路由、OSPF路由汇总、OSPF路由协议认证等配置与管理工作。

2.4 路由重分发与路由控制

问题描述

根据学校区网络架构，主校区网络采用动态路由协议OSPF，分校区网络采用动态路由协议RIP，所以需要采用路由重分发技术，实现全网互通。主校区与分校区出口默认路由以及接入层的直连路由也需要使用路由重分发技术，引入到主干网络中，实现网络互连互通。

分校区只能访问主校区的服务器群网络，不能访问别的区域，需要使用路由控制技术，对进出站路由进行控制，让路由器只学到必要的、可预知的路由。

问题分析

配置路由重分发与路由控制时，主要涉及RIP与OSPF路由重分发配置、直连路由重分发配置、静态路由重分发配置以及分发列表配置。

1. 配置RIP路由协议的重分发

步骤1 创建RIP路由进程。

router(config)#**router rip**

步骤2 配置RIP协议的路由重分发。Protocal表示路由重分发的源路由协议，metric metric-value表示重分发的路由度量值，match internal□external nssa-external type设置路由重分发的条件且只适合重分发的源路由协议为OSPF。router-map map-tag表示应用路由图进行重分发控制。

router(config-router)# redistribute protocal [metric metric-value][match internal□external nssa-external tyce][route-map map-tag]

2. 配置OSPF路由协议的重分发

步骤1 创建OSPF路由进程。

router(config)#**router ospf**

步骤2 配置OSPF协议的路由重分发。Protocal表示路由重分发的源路由协议，metric metric-value表示重分发的路由度量值，metric-type设置重分发的路由度量类型，tag tag-value表示重分发的路由的tag，route-map map-tag表示应用路由图进行重分发控制。

router(config-router)#**redistribute** protocal [**metric** metric-value] [**metric-type**{1｜2}][**tag** tag-value] [**router-map** map-tag]

3. 配置直连路由的重分发

（1）RIP协议重分发直连路由命令。

router(config-router)#**redistribute connected** [**metric** metric-value]

（2）OSPF协议重分发直连路由命令。

router(config-router)#**redistribute connected** [**subnets**] [**metric** metric-value] [**metric-type**{1｜2}][**tag** tag-value] [**router-map** map-tag]

4. 配置静态路由的重分发

（1）RIP协议重分发静态路由命令。

router(config-router)#**redistribute static** [**metric** metric-value]

（2）OSPF协议重分发静态路由命令。

router(config-router)#**redistribute static** [**subnets**] [**metric** metric-value] [**metric-type**{1｜2}][**tag** tag-value] [**router-map** map-tag]

5. 配置默认路由的重分发

（1）RIP协议重分发默认路由命令。

router(config-router)#**default-information originate** [**router-map** map-name]

（2）OSPF协议重分发默认路由命令。

router(config-router)#**default-information originate** [**always**] [**metric** metric-value] [**metric-type** type-value]

第2章 路由技术

[**router-map** *map-name*]

任 务 单

1	配置路由重分发
2	配置路由控制与过滤

根据任务单的安排完成任务。

▶ 任务1：配置路由重分发

任务实施

1. 任务描述及网络拓扑设计

由于RA路由器上运行RIP V2动态路由，RC路由器上运行OSPF动态路由，需要在运行不同路由协议的网络边界路由器RB上配置路由重分发，实现网络的互通。RB路由器的F0/0接口运行RIP V2路由协议，F0/1接口运行OSPF路由协议，在Area 0区域，L0和L1接口为外部路由；RC路由器的F0/1、L0和L1接口运行OSPF路由协议，其中F0/1接口在Area 0区域，L0和L1接口在Area 1区域。绘制拓扑结构图，如图2-9所示。

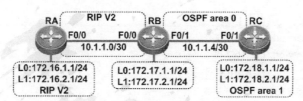

图2-9 配置路由重分发

2. 网络设备配置

（1）配置路由器各接口的IP地址

1）在RA路由器上配置IP地址。
RA#configure terminal
RA(config)#interface fastEthernet 0/0
RA(config-if-FastEthernet 0/0)#ip address 10.1.1.1 255.255.255.252
RA(config-if-FastEthernet 0/0)#exit
RA(config)#interface loopback 0
RA(config-if-Loopback 0)#ip address 172.16.1.1 255.255.255.0
RA(config-if-Loopback 0)#exit
RA(config)#interface loopback 1
RA(config-if-Loopback 1)#ip address 172.16.2.1 255.255.255.0
RA(config-if-Loopback 1)#exit

2）在RB路由器上配置IP地址。
RB(config)#interface fastEthernet 0/0
RB(config-if-FastEthernet 0/0)#ip address 10.1.1.2 255.255.255.252
RB(config-if-FastEthernet 0/0)#exit

RB(config)#interface fastEthernet 0/1
RB(config-if-FastEthernet 0/1)#ip address 10.1.1.5 255.255.255.252
RB(config-if-FastEthernet 0/1)#exit
RB(config)#interface loopback 0
RB(config-if-Loopback 0)#ip address 172.17.1.1 255.255.255.0
RB(config-if-Loopback 0)#exit
RB(config)#interface loopback 1
RB(config-if-Loopback 1)#ip address 172.17.2.1 255.255.255.0
RB(config-if-Loopback 1)#exit

3）在RC路由器上配置IP地址。
RC(config)#interface fastEthernet 0/1
RC(config-if-FastEthernet 0/1)#ip address 10.1.1.6 255.255.255.252
RC(config-if-FastEthernet 0/1)#exit
RC(config)#interface loopback 0
RC(config-if-Loopback 0)#ip address 172.18.1.1 255.255.255.0
RC(config-if-Loopback 0)#exit
RC(config)#interface loopback 1
RC(config-if-Loopback 1)#ip address 172.18.2.1 255.255.255.0

（2）配置RIP和OSPF
1）在RA路由器上配置RIP路由。
RA(config)#router rip
RA(config-router)#no auto-summary
RA(config-router)#version 2
RA(config-router)#network 10.1.1.0
RA(config-router)#network 172.16.1.0
RA(config-router)#network 172.16.2.0

2）在RB路由器上配置OSPF路由。
RB(config)#router ospf 100
RB(config-router)#network 10.1.1.4 0.0.0.3 area 0

3）在RB路由器上配置RIP路由。
RB(config)#router rip
RB(config-router)#version 2
RB(config-router)#network 10.1.1.0
RB(config-router)#no auto-summary

4）在RC路由器上配置OSPF路由。
RC(config)#router ospf 100
RC(config-router)#network 10.1.1.4 0.0.0.3 area 0
RC(config-router)#network 172.18.1.0 0.0.0.255 area 1
RC(config-router)#network 172.18.2.0 0.0.0.255 area 1

（3）配置路由重分发
1）在RB路由器上配置OSPF路由重分发。
RB(config)#router ospf 100
RB(config-router)#redistribute rip subnets   #将RIP路由重分发到OSPF中
RB(config-router)#redistribute connected subnets   #将直连路由重分发到OSPF中

2）在RB路由器上配置RIP路由重分发。
RB(config)#router rip
RB(config-router)#redistribute ospf 100 metric 1   #将OSPF路由重分发到RIP中

RB(config-router)#redistribute connected　　　　　　　#将直连路由重分发到RIP中

（4）验证测试

1）在RA路由器上查看路由表。

RA#show ip route

Codes:　C - connected, S - static, R - RIP, B - BGP
　　　　O - OSPF, IA - OSPF inter area
　　　　N1 - OSPF NSSA external type 1, N2 - OSPF NSSA external type 2
　　　　E1 - OSPF external type 1, E2 - OSPF external type 2
　　　　i - IS-IS, su - IS-IS summary, L1 - IS-IS level-1, L2 - IS-IS level-2
　　　　ia - IS-IS inter area, * - candidate default

Gateway of last resort is no set
C　　10.1.1.0/30 is directly connected, FastEthernet 0/0
C　　10.1.1.1/32 is local host.
R　　10.1.1.4/30 [120/1] via 10.1.1.2, 00:06:41, FastEthernet 0/0
C　　172.16.1.0/24 is directly connected, Loopback 0
C　　172.16.1.1/32 is local host.
C　　172.16.2.0/24 is directly connected, Loopback 1
C　　172.16.2.1/32 is local host.
R　　172.17.1.0/24 [120/1] via 10.1.1.2, 00:00:07, FastEthernet 0/0
R　　172.17.2.0/24 [120/1] via 10.1.1.2, 00:00:07, FastEthernet 0/0
R　　172.18.1.1/32 [120/1] via 10.1.1.2, 00:04:45, FastEthernet 0/0
R　　172.18.2.1/32 [120/1] via 10.1.1.2, 00:04:45, FastEthernet 0/0

从RA的路由表可以看出，RA学习到了被重分发的OSPF子网的路由信息。

2）在RC路由器上查看路由表。

RC#show ip route

Codes:　C - connected, S - static, R - RIP, B - BGP
　　　　O - OSPF, IA - OSPF inter area
　　　　N1 - OSPF NSSA external type 1, N2 - OSPF NSSA external type 2
　　　　E1 - OSPF external type 1, E2 - OSPF external type 2
　　　　i - IS-IS, su - IS-IS summary, L1 - IS-IS level-1, L2 - IS-IS level-2
　　　　ia - IS-IS inter area, * - candidate default

Gateway of last resort is no set
O E2 10.1.1.0/30 [110/20] via 10.1.1.5, 00:04:36, FastEthernet 0/1
C　　10.1.1.4/30 is directly connected, FastEthernet 0/1
C　　10.1.1.6/32 is local host.
O E2 172.16.1.0/24 [110/20] via 10.1.1.5, 00:04:36, FastEthernet 0/1
O E2 172.16.2.0/24 [110/20] via 10.1.1.5, 00:04:36, FastEthernet 0/1
O E2 172.17.1.0/24 [110/20] via 10.1.1.5, 00:00:21, FastEthernet 0/1
O E2 172.17.2.0/24 [110/20] via 10.1.1.5, 00:00:21, FastEthernet 0/1
C　　172.18.1.0/24 is directly connected, Loopback 0
C　　172.18.1.1/32 is local host.
C　　172.18.2.0/24 is directly connected, Loopback 1
C　　172.18.2.1/32 is local host.

从RC的路由表可以看出，RC学习到了被重分发的RIP子网的路由信息。

任务2：配置路由控制与过滤

任务实施

1. 任务描述及网络拓扑设计

由于RA路由器上运行OSPF动态路由，RC路由器上运行RIP V2动态路由，需要在运行不同路由协议的网络边界路由器RB上配置路由重分发，实现网络的互通。要求配置分发列表实现RA路由器L0接口子网信息可以发布给RIP路由，L1接口子网信息不发布给RIP路由。绘制拓扑结构图，如图2-10所示。

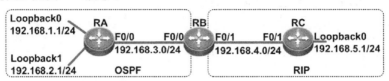

图2-10 配置路由控制与过滤

2. 网络设备配置

（1）配置路由器各接口的IP地址

1）在RA路由器上配置IP地址。
RA#configure terminal
RA(config)#interface fastEthernet 0/0
RA(config-if-FastEthernet 0/0)#ip address 192.168.3.1 255.255.255.0
RA(config-if-FastEthernet 0/0)#exit
RA(config)#interface loopback 0
RA(config-if-Loopback 0)#ip address 192.168.1.1 255.255.255.0
RA(config-if-Loopback 0)#exit
RA(config)#interface loopback 1
RA(config-if-Loopback 1)#ip address 192.168.2.1 255.255.255.0
RA(config-if-Loopback 1)#exit

2）在RB路由器上配置IP地址。
RB(config)#interface fastEthernet 0/0
RB(config-if-FastEthernet 0/0)#ip address 192.168.3.2 255.255.255.0
RB(config-if-FastEthernet 0/0)#exit
RB(config)#interface fastEthernet 0/1
RB(config-if-FastEthernet 0/1)#ip address 192.168.4.1 255.255.255.0
RB(config-if-FastEthernet 0/1)#exit

3）在RC路由器上配置IP地址。
RC(config)#interface fastEthernet 0/1
RC(config-if-FastEthernet 0/1)#ip address 192.168.4.2 255.255.255.0
RC(config-if-FastEthernet 0/1)#exit
RC(config)#interface loopback 0
RC(config-if-Loopback 0)#ip address 192.168.5.1 255.255.255.0
RC(config-if-Loopback 0)#exit

（2）配置RIP和OSPF路由

1）在RA路由器上配置OSPF路由。
RA(config)#router ospf 100

```
RA(config-router)#network 192.168.3.0 0.0.0.255 area 0
RA(config-router)#network 192.168.1.0 0.0.0.255 area 0
RA(config-router)#network 192.168.2.0 0.0.0.255 area 0
RA(config-router)#exit
```

2）在RB路由器上配置OSPF路由。
```
RB(config)#router ospf 100
RB(config-router)#network 192.168.3.0 0.0.0.255 area 0
RB(config-router)#exit
```

3）在RB路由器上配置RIP路由。
```
RB(config)#router rip
RB(config-router)#no auto-summary
RB(config-router)#version 2
RB(config-router)#network 192.168.4.0
RB(config-router)#exit
```

4）在RC路由器上配置RIP路由。
```
RC(config)#router rip
RC(config-router)#no auto-summary
RC(config-router)#version 2
RC(config-router)#network 192.168.5.0
RC(config-router)#network 192.168.4.0
RC(config-router)#exit
```

(3) 在RB路由器上配置路由重分发
```
RB(config)#router ospf 100
RB(config-router)#redistribute rip subnets              #将RIP路由重分发到OSPF中
RB(config-router)#redistribute connected subnets        #将直连路由重分发到OSPF中
RB(config-router)#exit
RB(config)#router rip
RB(config-router)#redistribute ospf 100 metric 1        #将OSPF路由重分发到RIP中
RB(config-router)#redistribute connected               #将直连路由重分发到RIP中
RB(config-router)#exit
```

(4) 在RB路由器上配置分发列表
```
RB(config)#ip access-list standard 10                   #配置访问控制列表
RB(config-std-nacl)#deny 192.168.1.0 0.0.0.255          #拒绝192.168.1.0网段
RB(config-std-nacl)#permit any                          #允许所有网段
RB(config-std-nacl)#exit
RB(config)#router rip
RB(config-router)#distribute-list 10 out fastEthernet 0/1   #配置分发列表
RB(config-router)#exit
```

(5) 验证测试
```
RC#show ip route

Codes:  C - connected, S - static, R - RIP, B - BGP
        O - OSPF, IA - OSPF inter area
        N1 - OSPF NSSA external type 1, N2 - OSPF NSSA external type 2
        E1 - OSPF external type 1, E2 - OSPF external type 2
        i - IS-IS, su - IS-IS summary, L1 - IS-IS level-1, L2 - IS-IS level-2
        ia - IS-IS inter area, * - candidate default

Gateway of last resort is no set
```

R   192.168.2.1/32 [120/1] via 192.168.4.1, 00:04:03, FastEthernet 0/1
R   192.168.3.0/24 [120/1] via 192.168.4.1, 00:04:03, FastEthernet 0/1
C   192.168.4.0/24 is directly connected, FastEthernet 0/1
C   192.168.4.2/32 is local host.
C   192.168.5.0/24 is directly connected, Loopback 0
C   192.168.5.1/32 is local host.

从输出结果可以看到，在配置了分发列表后，路由器RC无法学习到RA路由器的L1接口192.168.1.0/24网段信息。

小结

通过路由重分发和路由控制的学习，主要掌握RIP与OSPF路由重分发、静态路由重分发、直连路由重分发、默认路由重分发、配置分发列表、调整AD值等配置与管理工作。

2.5 策略路由

问题描述

主校区采用双出口网络架构，为了区分不同VLAN用户访问互联网的流量，设置部分内网用户使用不同的出口访问互联网。同时要求部分内网用户访问教育网资源时走教育网出口，访问其他资源时走电信出口。

问题分析

配置策略路由时，主要涉及基于源地址的策略路由配置、基于目的地址的策略路由配置、基于报文长度的策略路由配置。

配置基于策略的路由选择步骤如下：

步骤1   配置route-map命令。route-map-name是route-map的名称，permit表示如果报文符合该子句中的匹配条件，则报文将被进行策略路由，deny表示如果报文符合该子句中的匹配条件，则报文将不进行策略路由，而进行正常的转发，sequence-number是route-map子句的编号，route-map中各子句按照编号的顺序执行。

　　router(config)#**route-map** *route-map-name* **[permit | deny]**[*sequence-number*]

步骤2   配置match命令。access-list-number | name表示访问控制列表编号或者名称，用于匹配入站报文，min表示最小报文长度，max表示最大报文长度。

　　router(config-route-map)#**match ip address** {*access-list-number | name*}
　　router(config-route-map)#**match length** *min max*

步骤3   配置set命令。ip-address用于指定报文前往目的地路径中相邻下一跳路由器的地址，interface用于指定报文被转发的本地出口。

第2章 路由技术

```
router(config-route-map)#set ip next-hop {ip-address}
router(config-route-map)#set interface interface
```

步骤4 在接口上配置策略路由。

```
router(config-if)#ip policy route-map route-map-name
```

任 务 单

1	配置基于源地址的策略路由
2	配置基于目的地址的策略路由
3	配置基于报文长度的策略路由

根据任务单的安排完成任务。

任务1：配置基于源地址的策略路由

任务实施

1. 任务描述及网络拓扑设计

在路由器RA上配置基于源地址的策略路由，实现将源地址10.1.1.10/24的用户报文转发到路由器RB的F0/0接口，将源地址为10.1.1.20/24的用户报文转发到路由器RC的F0/0接口。绘制拓扑结构图，如图2-11所示。

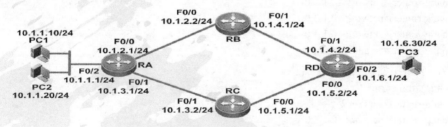

图2-11 配置基于源地址的策略路由

2. 网络设备配置

（1）配置路由器各接口的IP地址

1）在RA路由器上配置IP地址。

```
RA(config)#interface fastEthernet 0/0
RA(config-if-FastEthernet 0/0)#ip address 10.1.2.1 255.255.255.0
RA(config-if-FastEthernet 0/0)#exit
RA(config)#interface fastEthernet 0/1
RA(config-if-FastEthernet 0/1)#ip address 10.1.3.1 255.255.255.0
RA(config-if-FastEthernet 0/1)#exit
RA(config)#interface fastEthernet 0/2
RA(config-if-FastEthernet 0/2)#ip address 10.1.1.1 255.255.255.0
RA(config-if-FastEthernet 0/2)#exit
```

2）在RB路由器上配置IP地址。

RB(config)#interface fastEthernet 0/0
RB(config-if-FastEthernet 0/0)#ip address 10.1.2.2 255.255.255.0
RB(config-if-FastEthernet 0/0)#exit
RB(config)#interface fastEthernet 0/1
RB(config-if-FastEthernet 0/1)#ip address 10.1.4.1 255.255.255.0
RB(config-if-FastEthernet 0/1)#exit

3）在RC路由器上配置IP地址。
RC(config)#interface fastEthernet 0/1
RC(config-if-FastEthernet 0/1)#ip address 10.1.3.2 255.255.255.0
RC(config-if-FastEthernet 0/1)#exit
RC(config)#interface fastEthernet 0/0
RC(config-if-FastEthernet 0/0)#ip address 10.1.5.1 255.255.255.0

4）在RD路由器上配置IP地址。
RD(config)#interface fastEthernet 0/1
RD(config-if-FastEthernet 0/1)#ip address 10.1.4.2 255.255.255.0
RD(config-if-FastEthernet 0/1)#exit
RD(config)#interface fastEthernet 0/0
RD(config-if-FastEthernet 0/0)#ip address 10.1.5.2 255.255.255.0
RD(config-if-FastEthernet 0/0)#exit
RD(config)#interface fastEthernet 0/2
RD(config-if-FastEthernet 0/2)#ip address 10.1.6.1 255.255.255.0

（2）配置RIP路由

1）在RA路由器上配置RIP路由。
RA(config)#router rip
RA(config-router)#version 2
RA(config-router)#network 10.1.1.0
RA(config-router)#network 10.1.2.0
RA(config-router)#network 10.1.3.0
RA(config-router)#no auto-summary

2）在RB路由器上配置RIP路由。
RB(config)#router rip
RB(config-router)#version 2
RB(config-router)#network 10.1.2.0
RB(config-router)#network 10.1.4.0
RB(config-router)#no auto-summary

3）在RC路由器上配置RIP路由。
RC(config)#router rip
RC(config-router)#version 2
RC(config-router)#network 10.1.3.0
RC(config-router)#network 10.1.5.0
RC(config-router)#no auto-summary

4）在RD路由器上配置RIP路由。
RD(config)#router rip
RD(config-router)#version 2
RD(config-router)#network 10.1.4.0
RD(config-router)#network 10.1.5.0
RD(config-router)#network 10.1.6.0
RD(config-router)#no auto-summary

(3) 配置PBR

```
RA(config)#ip access-list standard 10
RA(config-std-nacl)#permit host 10.1.1.10
RA(config-std-nacl)#exit
RA(config)#ip access-list standard 11
RA(config-std-nacl)#permit host 10.1.1.20
RA(config-std-nacl)#exit
RA(config)#route-map source permit 10          #配置名为source的route-map
RA(config-route-map)#match ip address 10
        #匹配ACL名称为10的数据执行下面的动作
RA(config-route-map)#set ip next-hop 10.1.2.2  #设置下一跳地址为10.1.2.2
RA(config-route-map)#exit
RA(config)#route-map source permit 20          #配置名为source的route-map
RA(config-route-map)#match ip address 11
        #匹配ACL名称为11的数据执行下面的动作
RA(config-route-map)#set ip next-hop 10.1.3.2  #设置下一跳地址为10.1.3.2
```

(4) 在报文的入站接口应用router-map

```
RA(config)#interface fastEthernet 0/2
RA(config-if-FastEthernet 0/2)#ip policy route-map source    #应用route-map
```

(5) 验证测试

1) 设置主机PC1的IP地址为10.1.1.10/24,在主机PC1上用tracert命令测试数据包发送路径,如图2-12所示。

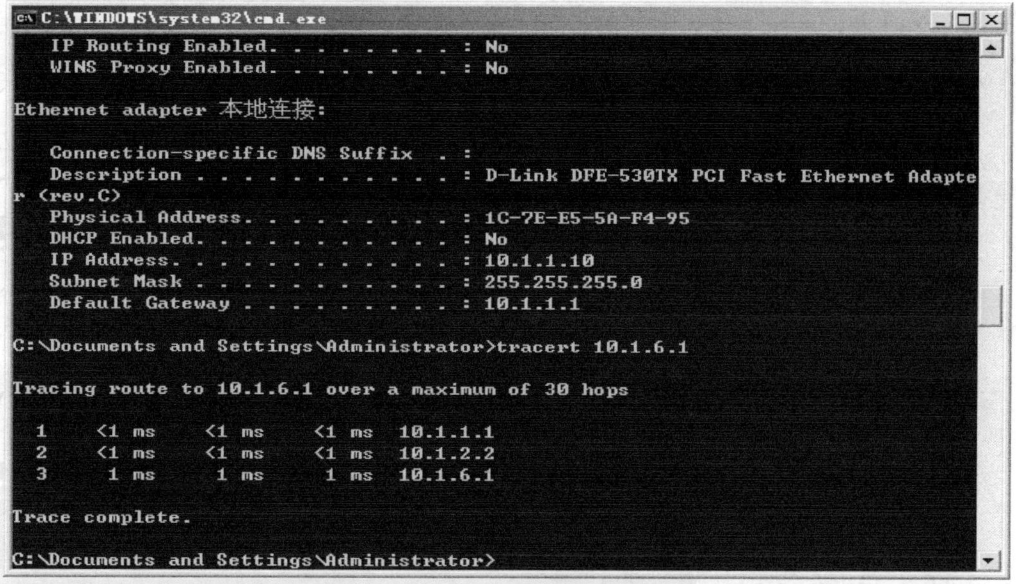

```
C:\WINDOWS\system32\cmd.exe

        IP Routing Enabled. . . . . . . . : No
        WINS Proxy Enabled. . . . . . . . : No

Ethernet adapter 本地连接:

        Connection-specific DNS Suffix  . :
        Description . . . . . . . . . . . : D-Link DFE-530TX PCI Fast Ethernet Adapte
r (rev.C)
        Physical Address. . . . . . . . . : 1C-7E-E5-5A-F4-95
        DHCP Enabled. . . . . . . . . . . : No
        IP Address. . . . . . . . . . . . : 10.1.1.10
        Subnet Mask . . . . . . . . . . . : 255.255.255.0
        Default Gateway . . . . . . . . . : 10.1.1.1

C:\Documents and Settings\Administrator>tracert 10.1.6.1

Tracing route to 10.1.6.1 over a maximum of 30 hops

  1    <1 ms    <1 ms    <1 ms  10.1.1.1
  2    <1 ms    <1 ms    <1 ms  10.1.2.2
  3     1 ms     1 ms     1 ms  10.1.6.1

Trace complete.

C:\Documents and Settings\Administrator>
```

图2-12 设置主机PC1的IP地址 (1)

从tracert结果可以看到,主机PC1发送的数据包通过路由器RB进行转发。

2) 设置主机PC2的IP地址为10.1.1.20/24,在主机PC2上用tracert命令测试数据包发送路径,如图2-13所示。

从tracert结果可以看到,主机PC2发送的数据包通过路由器RC进行转发。

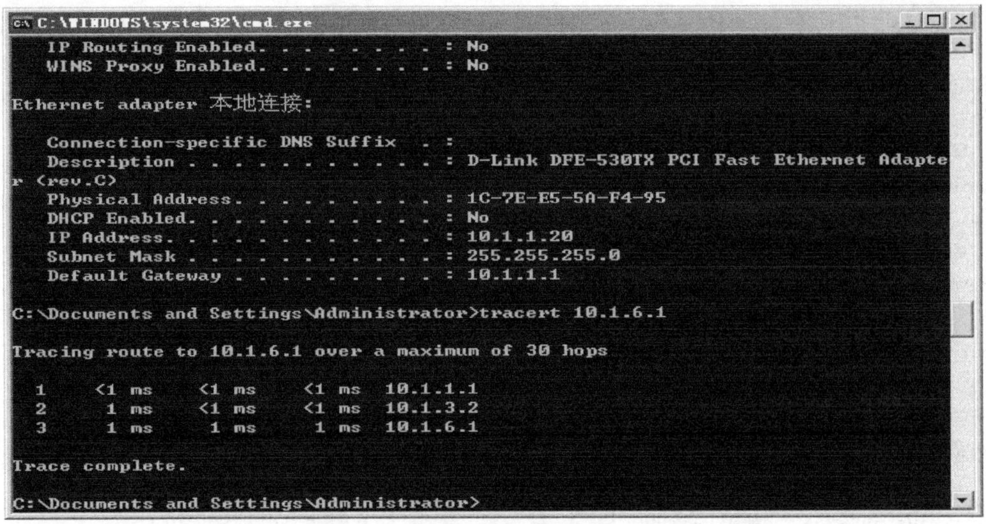

图2-13 设置主机PC2的IP地址（1）

任务2：配置基于目的地址的策略路由

任务实施

1. 任务描述及网络拓扑设计

在路由器RA上配置基于目的地址策略路由，当10.1.1.0/24网段的主机访问目的地址为10.1.6.100/24主机时，报文转发到路由器RB的F0/0接口；当10.1.1.0/24网段的主机访问目的地址为10.1.6.200/24主机时，报文转发到路由器RC的F0/1接口。绘制拓扑结构图，如图2-14所示。

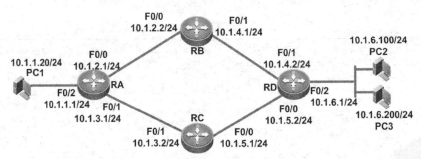

图2-14 配置基于目的地址的策略路由

2. 网络设备配置

（1）配置路由器各接口的IP地址

1）在RA路由器上配置IP地址。

RA(config)#interface fastEthernet 0/0
RA(config-if-FastEthernet 0/0)#ip address 10.1.2.1 255.255.255.0
RA(config-if-FastEthernet 0/0)#exit
RA(config)#interface fastEthernet 0/1
RA(config-if-FastEthernet 0/1)#ip address 10.1.3.1 255.255.255.0

```
RA(config-if-FastEthernet 0/1)#exit
RA(config)#interface fastEthernet 0/2
RA(config-if-FastEthernet 0/2)#ip address 10.1.1.1 255.255.255.0
RA(config-if-FastEthernet 0/2)#exit
```

2）在RB路由器上配置IP地址。
```
RB(config)#interface fastEthernet 0/0
RB(config-if-FastEthernet 0/0)#ip address 10.1.2.2 255.255.255.0
RB(config-if-FastEthernet 0/0)#exit
RB(config)#interface fastEthernet 0/1
RB(config-if-FastEthernet 0/1)#ip address 10.1.4.1 255.255.255.0
RB(config-if-FastEthernet 0/1)#exit
```

3）在RC路由器上配置IP地址。
```
RC(config)#interface fastEthernet 0/1
RC(config-if-FastEthernet 0/1)#ip address 10.1.3.2 255.255.255.0
RC(config-if-FastEthernet 0/1)#exit
RC(config)#interface fastEthernet 0/0
RC(config-if-FastEthernet 0/0)#ip address 10.1.5.1 255.255.255.0
RC(config-if-FastEthernet 0/0)#exit
```

4）在RD路由器上配置IP地址。
```
RD(config)#interface fastEthernet 0/1
RD(config-if-FastEthernet 0/1)#ip address 10.1.4.2 255.255.255.0
RD(config-if-FastEthernet 0/1)#exit
RD(config)#interface fastEthernet 0/0
RD(config-if-FastEthernet 0/0)#ip address 10.1.5.2 255.255.255.0
RD(config-if-FastEthernet 0/0)#exit
RD(config)#interface fastEthernet 0/2
RD(config-if-FastEthernet 0/2)#ip address 10.1.6.1 255.255.255.0
RD(config-if-FastEthernet 0/2)#exit
```

（2）在路由器上配置RIP路由

1）在RA路由器上配置RIP路由。
```
RA(config)#router rip
RA(config-router)#no auto-summary
RA(config-router)#version 2
RA(config-router)#network 10.1.1.0
RA(config-router)#network 10.1.2.0
RA(config-router)#network 10.1.3.0
RA(config-router)#exit
```

2）在RB路由器上配置RIP路由。
```
RB(config)#router rip
RB(config-router)#version 2
RB(config-router)#network 10.1.2.0
RB(config-router)#network 10.1.4.0
RB(config-router)#no auto-summary
```

3）在RC路由器上配置RIP路由。
```
RC(config)#router rip
RC(config-router)#version 2
RC(config-router)#network 10.1.3.0
RC(config-router)#network 10.1.5.0
```

RC(config-router)#no auto-summary

4）在RD路由器上配置RIP路由。
RD(config)#router rip
RD(config-router)#version 2
RD(config-router)#network 10.1.4.0
RD(config-router)#network 10.1.5.0
RD(config-router)#network 10.1.6.0
RD(config-router)#no auto-summary

（3）配置PBR
RA(config)#ip access-list extended 100
RA(config-ext-nacl)#permit ip any host 10.1.6.100
RA(config-ext-nacl)#exit
RA(config)#ip access-list extended 101
RA(config-ext-nacl)#permit ip any host 10.1.6.200
RA(config-ext-nacl)#exit
RA(config)#route-map des permit 10   #配置名为des的route-map
RA(config-route-map)#match ip address 100
         #匹配access-list 100的数据执行下面的动作
RA(config-route-map)#set ip next-hop 10.1.2.2   #设置下一跳地址为10.1.2.2
RA(config-route-map)#exit
RA(config)#route-map des permit 20   #配置名为des的route-map
RA(config-route-map)#match ip address 101
         #匹配access-list 101的数据执行下面的动作
RA(config-route-map)#set ip next-hop 10.1.3.2   #设置下一跳地址为10.1.3.2
RA(config-route-map)#exit

（4）在报文的入站接口应用router-map
RA(config)#interface fastEthernet 0/2
RA(config-if-FastEthernet 0/2)#ip policy route-map des   #应用route-map
RA(config-if-FastEthernet 0/2)#exit

（5）验证测试
　　1）设置主机PC1的IP地址为10.1.1.20/24，在主机PC1上用tracert命令进行路由跟踪，当目标地址为10.1.6.100时，测试如图2-15所示。

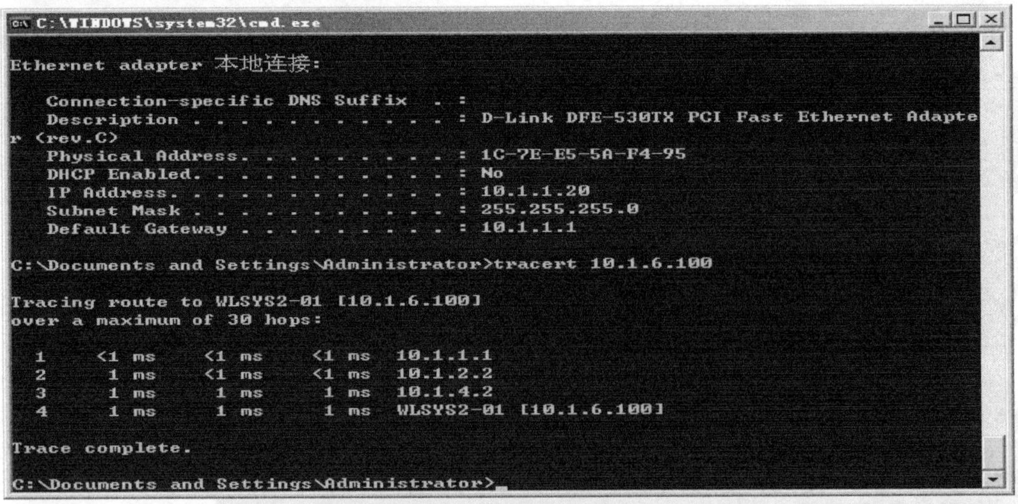

图2-15　设置主机PC1的IP地址（2）

从tracert结果可以看到，当数据包的目的地址是10.1.6.100时，数据包从RB路由器转发。

2）设置主机PC2的IP地址为10.1.1.20/24，在主机PC2上用tracert命令进行路由跟踪，当目标地址为10.1.6.200时，测试如图2-16所示。

```
C:\WINDOWS\system32\cmd.exe

Ethernet adapter 本地连接：

    Connection-specific DNS Suffix  . :
    Description . . . . . . . . . . . : D-Link DFE-530TX PCI Fast Ethernet Adapter (rev.C)
    Physical Address. . . . . . . . . : 1C-7E-E5-5A-F4-95
    DHCP Enabled. . . . . . . . . . . : No
    IP Address. . . . . . . . . . . . : 10.1.1.20
    Subnet Mask . . . . . . . . . . . : 255.255.255.0
    Default Gateway . . . . . . . . . : 10.1.1.1

C:\Documents and Settings\Administrator>tracert 10.1.6.200

Tracing route to WLSYS2-01 [10.1.6.200]
over a maximum of 30 hops:

  1    <1 ms    <1 ms    <1 ms  10.1.1.1
  2     1 ms    <1 ms    <1 ms  10.1.3.2
  3     1 ms     1 ms     2 ms  10.1.5.2
  4     2 ms     1 ms     1 ms  WLSYS2-01 [10.1.6.200]

Trace complete.

C:\Documents and Settings\Administrator>
```

图2-16 设置主机PC2的IP地址（2）

从tracert结果可以看到，当数据包的目的地址是10.1.6.200时，数据包从RC路由器转发。

任务3：配置基于报文长度的策略路由

任务实施

1. 任务描述及网络拓扑设计

在路由器RA上配置基于报文长度的策略路由，当10.1.1.0/24网段的主机访问外网的报文长度在150～1500B时，通过路由器RB；当10.1.1.0/24网段的主机访问外网的报文长度小于150B时，通过路由器RC。绘制拓扑结构图，如图2-17所示。

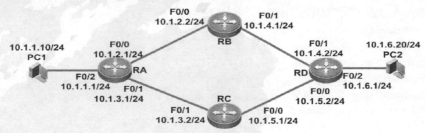

图2-17 配置基于报文长度的策略路由

2. 网络设备配置

（1）配置路由器各接口的IP地址

1）在RA路由器上配置IP地址。

RA(config)#interface fastEthernet 0/0
RA(config-if-FastEthernet 0/0)#ip address 10.1.2.1 255.255.255.0
RA(config-if-FastEthernet 0/0)#exit

RA(config)#interface fastEthernet 0/1
RA(config-if-FastEthernet 0/1)#ip address 10.1.3.1 255.255.255.0
RA(config-if-FastEthernet 0/1)#exit
RA(config)#interface fastEthernet 0/2
RA(config-if-FastEthernet 0/2)#ip address 10.1.1.1 255.255.255.0
RA(config-if-FastEthernet 0/2)#exit

2）在RB路由器上配置IP地址。
RB(config)#interface fastEthernet 0/0
RB(config-if-FastEthernet 0/0)#ip address 10.1.2.2 255.255.255.0
RB(config-if-FastEthernet 0/0)#exit
RB(config)#interface fastEthernet 0/1
RB(config-if-FastEthernet 0/1)#ip address 10.1.4.1 255.255.255.0
RB(config-if-FastEthernet 0/1)#exit

3）在RC路由器上配置IP地址。
RC(config)#interface fastEthernet 0/1
RC(config-if-FastEthernet 0/1)#ip address 10.1.3.2 255.255.255.0
RC(config-if-FastEthernet 0/1)#exit
RC(config)#interface fastEthernet 0/0
RC(config-if-FastEthernet 0/0)#ip address 10.1.5.1 255.255.255.0
RC(config-if-FastEthernet 0/0)#exit

4）在RD路由器上配置IP地址。
RD(config)#interface fastEthernet 0/1
RD(config-if-FastEthernet 0/1)#ip address 10.1.4.2 255.255.255.0
RD(config-if-FastEthernet 0/1)#exit
RD(config)#interface fastEthernet 0/0
RD(config-if-FastEthernet 0/0)#ip address 10.1.5.2 255.255.255.0
RD(config-if-FastEthernet 0/0)#exit
RD(config)#interface fastEthernet 0/2
RD(config-if-FastEthernet 0/2)#ip address 10.1.6.1 255.255.255.0
RD(config-if-FastEthernet 0/2)#exit

（2）配置RIP路由

1）在RA路由器上配置RIP路由。
RA(config)#router rip
RA(config-router)#version 2
RA(config-router)#network 10.1.1.0
RA(config-router)#network 10.1.2.0
RA(config-router)#network 10.1.3.0
RA(config-router)#no auto-summary
RA(config-router)#exit

2）在RB路由器上配置RIP路由。
RB(config)#router rip
RB(config-router)#version 2
RB(config-router)#network 10.1.2.0
RB(config-router)#network 10.1.4.0
RB(config-router)#no auto-summary
RB(config-router)#exit

3）在RC路由器上配置RIP路由。
RC(config)#router rip

```
RC(config-router)#version 2
RC(config-router)#network 10.1.3.0
RC(config-router)#network 10.1.5.0
RC(config-router)#no auto-summary
RC(config-router)#exit
```

4）在RD路由器上配置RIP路由。

```
RD(config)#router rip
RD(config-router)#version 2
RD(config-router)#network 10.1.4.0
RD(config-router)#network 10.1.5.0
RD(config-router)#network 10.1.6.0
RD(config-router)#no auto-summary
RD(config-router)#exit
```

（3）配置策略路由

```
RA(config)#route-map aaa permit 10
        #配置名为aaa的route-map
RA(config-route-map)#match length 150 1500
        #配置报文长度在150～1500B的匹配规则
RA(config-route-map)#set ip next-hop 10.1.2.2
        #配置报文长度在150～1500B的报文下一条地址为10.1.2.2
RA(config-route-map)#exit
RA(config)#route-map aaa permit 20
        #配置名为aaa的route-map
RA(config-route-map)#match length 0 150
        #配置报文长度小于150B的匹配规则
RA(config-route-map)#set ip next-hop 10.1.3.2
        #配置报文长度小于150B的报文下一条地址为10.1.3.2
RA(config-route-map)#exit
```

（4）在报文的入站接口上应用router-map

```
RA(config)#interface fastEthernet 0/2
RA(config-if-FastEthernet 0/2)#ip policy route-map aaa
        #在接口上应用名称为aaa的route-map
```

小结

通过策略路由的学习，主要掌握基于源地址的策略路由、基于目的地址的策略路由和基于报文长度的策略路由等配置与管理工作。

只有将router-map应用在报文的入站接口上，PBR才会生效。

第3章 网络安全技术

目前,最常用的局域网技术就是以太网技术。众所周知,以太网是一种广播介质,即网络中一个节点发送的数据都有可能会被其他节点所接收,攻击者只要接入到网络中,就很容易侦听到网络中发送的信息,这可以说是以太网先天性的缺陷。此外,在局域网中使用的各种协议、技术等都存在着安全隐患,如ARP、STP等,这些协议的开发者在开发协议时并没有考虑到安全因素,因此导致了协议能够被攻击者利用,产生网络攻击。

据相关数据显示,目前存在的网络安全威胁60%以上都是来自于内部网络,也可以说来自于局域网,例如ARP攻击、MAC地址欺骗、针对DHCP的DoS攻击等,而且这种内部的攻击所造成的损失也是巨大的。

从另一个角度来看,局域网中数据转发的设备主要为传统的交换机、路由器,这些设备的默认策略都是对所有数据进行转发的,并且没有启用任何安全机制,这也需要我们在这些设备上添加各种安全机制,以防范各种网络威胁。

本章介绍的网络安全控制技术主要涉及端口安全、DHCP监听、ARP检查、ACL应用和网络地址转换技术。

1. 端口安全

交换机的端口安全机制是工作在交换机二层端口上的一个安全特性。利用端口安全这个特性,可以实现网络接入安全,具体可以通过限制允许访问交换机上某个端口的MAC地址以及IP(可选)来实现对该端口输入的严格控制。

2. DHCP监听

DHCP监听能够通过过滤网络中接入的伪DHCP服务器发送的DHCP报文增强网络安全性。DHCP监听还可以检查DHCP客户端发送的DHCP报文的合法性,防止DHCP DoS攻击。

3. ARP检查

ARP是局域网中用来进行地址解析的协议,它将IP地址映射到MAC地址。ARP检查特性是交换机中防范ARP欺骗攻击的一个安全特性。

ARP检查特性的实现依赖于端口安全特性,也就是说,要使交换机的端口具有防范ARP欺骗的功能,首先要启用端口安全特性。ARP检查特性是通过查看端口所收到的ARP报文中嵌入的IP地址是否与配置的安全地址符合,如果不符合则将其视为非法的ARP报文。

动态ARP检查是交换机中用来防止ARP欺骗攻击的一种安全特性。DAI部署前提是需要DHCP环境的支持,也就是网络中客户端的IP地址是通过DHCP来进行的,而且DAI需要依据DHCP监听特性。

第3章 网络安全技术

4. ACL应用

ACL应用的主要目的是对网络数据通信进行过滤,从而实现各种访问控制需求。ACL技术通过对数据包中的五元组(源地址、目的地址、协议号、源端口号、目标端口号)来区分特定的数据流,并对匹配预设规则的数据采取相应的措施,允许或拒绝数据通过,从而实现对网络的安全控制。

5. 网络地址转换技术

NAT是将IP报头中的源地址或目的地址进行翻译或转换的一种技术,主要被用来将内部私有地址转换为公有地址。

3.1 端口安全

问题描述

学校校园网络规划中,部分区域要求只允许特定MAC地址的设备接入到网络中,从而防止用户将非法设备接入到网络中,所以在接入层安全方面,需要部署端口安全技术,对所有接入端口配置端口安全,限制端口接入数量为1个主机。如果有违规者,则关闭端口,为了保证接入层网络收敛的速度,在接入层上接口需要配置为速端口。

问题分析

交换机的端口安全机制是工作在交换机二层端口上的一个安全特性。交换机端口安全主要有以下几个功能:

1)只允许特定MAC地址的设备接入到网络中,从而防止用户将非法或未授权的设备接入网络。

2)限制端口接入的设备数量,防止用户将过多的设备接入到网络中。

交换机端口安全配置时,主要涉及配置安全地址个数、违规策略、地址绑定等。

配置端口安全步骤如下:

步骤1 开启端口安全。

switch(config-if)#**switchport port-security**

步骤2 配置最大安全地址个数。number表示最大安全地址个数。

switch(config-if)#**switchport port-security maximum** *number*

步骤3 配置静态安全地址绑定。mac-address表示绑定的MAC地址。

switch(config-if)#**switchport port-security mac-address** *mac-address*

步骤4 配置地址老化时间。如果此命令指定了关键字static,那么老化时间也将应用到手工配置的安全地址。默认情况下,老化时间只应用于动态学习的安全地址,手工配置的安全地址是永远存在的。

switch(config-if)#**switchport port-security aging {time** *time* **| static}**

步骤5 配置地址违规的操作行为。protect关键字表示地址违规发生时，交换机将丢弃接收到的帧，但是交换机将不会通知违规的产生；restrict关键字表示地址违规发生时，交换机不但丢弃接收到的帧，而且发送一个SNMP Trap报文；shotdown关键字表示地址违规发生时，交换机将丢弃接收到的帧，发送一个SNMP Trap报文，而且将关闭端口。

switch(config-if)#**switchport port-security violation {protect | restrict | shutdown}**

任 务 单

1	配置交换机端口安全

解决步骤

根据任务单的安排完成任务。

任务：配置交换机端口安全

任务实施

1. 任务描述及网络拓扑设计

在交换机上启用端口安全特性，配置PC1的MAC地址为安全地址，PC1所连接的F0/1端口最大安全地址个数为1，当产生地址违例时关闭端口。绘制拓扑结构图，如图3-1所示。

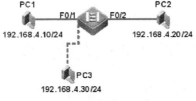

图3-1 配置端口安全

2. 网络设备配置

（1）启用端口安全特性

switch>enable
switch#configure terminal
switch(config)#interface fastEthernet 0/1
switch(config-FastEthernet 0/1)#switchport port-security   #启用端口安全

（2）配置PC1的MAC地址为安全地址

switch(config-FastEthernet 0/1)#switchport port-security mac-address 0001.0001.0001
                                        #配置安全地址

（3）配置最多允许一个安全地址，即保证只有PC1可以接入到此端口

switch(config-FastEthernet 0/1)#switchport port-security maximum 1
                                        #允许最大连接个数为1

（4）配置违例产生方式

switch(config-FastEthernet 0/1)#switchport port-security violation shutdown
                                        #如果违规则关闭端口

（5）验证测试

将PC3机器接入到F0/1端口，并且设置IP地址为192.168.4.30。在PC3上ping PC2的IP地址，无法ping通，交换机会关闭端口，出现下列违例提示。

Ruijie(config)#*Jan　8 15:23:55: %LINEPROTO-5-UPDOWN: Line protocol on Interface FastEthernet 0/1, changed state to up.
　*Jan　8 15:23:56: %PORT_SECURITY-2-PSECURE_VIOLATION: Security violation occurred, caused by MAC address 1c7e.e55a.f4a4 on port FastEthernet 0/1.
　*Jan　8 15:23:58: %LINK-3-UPDOWN: Interface FastEthernet 0/1, changed state to down.
　*Jan　8 15:23:58: %LINEPROTO-5-UPDOWN: Line protocol on Interface FastEthernet 0/1, changed state to down.

由于PC3的MAC地址不是所配置的安全地址，而且由于端口最多允许1个安全MAC地址，所以当端口收到PC3发送的数据帧时，产生了端口违规现象，端口被关闭。

小结

通过端口安全的学习，主要掌握端口安全的默认配置及违规产生方式、端口安全地址、安全地址老化时间等配置与管理工作。

3.2　DHCP监听

问题描述

在主校区信息中心服务器群中，部署了DHCP服务器，为内网用户分配IP地址，网络中会存在DHCP DoS攻击和DHCP无赖设备的攻击，为了保障DHCP服务器正常工作，需要在网络中采用DHCP监听功能和DHCP中继功能。

问题分析

DHCP监听能够通过过滤网络中接入的伪DHCP服务器报文增强网络安全性。DHCP监听还可以检查DHCP客户端发送的DHCP报文的合法性，防止DHCP DoS攻击。

配置DHCP监听时，主要涉及配置有开启DHCP监听功能、设置信任端口、配置DHCP监听绑定表项、将DHCP监听绑定表项写入数据库等。

配置DHCP监听步骤如下：

步骤1　开启DHCP监听功能。

switch(config)#**ip dhcp snooping**

步骤2　设置DHCP监听信任端口。需要将DHCP服务器的端口与交换机上行链路端口设置为信任端口。

switch(config-if)#**ip dhcp snooping trust**

步骤3 配置DHCP监听绑定表项。

switch(config-if)#**ip dhcp snooping binding** *mac-address* **vlan** *vlan-id* **ip** *ip-address* **interface** *interface*

步骤4 手工将DHCP监听绑定表项写入数据库。

switch(config)#**ip dhcp snooping database write-to-flash**

步骤5 系统自动将DHCP监听绑定表项写入数据库。

switch(config)#**ip dhcp snooping database write-delay** *seconds*

任 务 单

1	配置DHCP监听

解决步骤

根据任务单的安排完成任务。

任务：配置DHCP监听

任务实施

1. 任务描述及网络拓扑设计

配置两台DHCP服务器，VLAN 10中的服务器为伪服务器，VLAN 20中的服务器为合法服务器。伪服务器中的地址池为1.1.1.0/24，合法服务器中的地址池为10.1.10.0/24。绘制拓扑结构图，如图3-2所示。

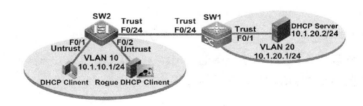

图3-2　DHCP监听

2. 网络设备配置

（1）在SW1和SW2交换机上配置VLAN及端口划分

1）在SW1交换机上配置VLAN及端口划分。

SW1(config)#vlan 10
SW1(config-vlan)#exit
SW1(config)#vlan 20
SW1(config-vlan)#exit
SW1(config)#interface fastEthernet 0/1
SW1(config-FastEthernet 0/1)#switchport access vlan 20
SW1(config-FastEthernet 0/1)#exit
SW1(config)#interface fastEthernet 0/24
SW1(config-FastEthernet 0/24)#switchport mode trunk
SW1(config-FastEthernet 0/24)#exit

```
SW1(config)#interface vlan 10
SW1(config-VLAN 10)#ip address 10.1.10.1 255.255.255.0
SW1(config-VLAN 10)#exit
SW1(config)#interface vlan 20
SW1(config-VLAN 20)#ip address 10.1.20.1 255.255.255.0
SW1(config-VLAN 20)#exit
```

2）在SW2交换机上配置VLAN及端口划分。

```
SW2(config)#vlan 10
SW2(config-vlan)#exit
SW2(config)#interface range fastEthernet 0/1-2
SW2(config-if-range)#switchport access vlan 10
SW2(config-if-range)#exit
SW2(config)#interface fastEthernet 0/24
SW2(config-FastEthernet 0/24)#switchport mode trunk
SW2(config-FastEthernet 0/24)#exit
```

（2）在SW1上配置DHCP Relay

```
SW1(config)#service dhcp                              #启用DHCP服务
SW1(config)#interface vlan 10                         #进入VLAN接口模式
SW1(config-VLAN 10)# ip help-address 10.1.20.2        #配置DHCP中继
```

（3）验证测试

确保服务器可以正常工作。将客户端PC配置为自动获取地址后，接入交换机SW2的F0/2端口后，此时可以看到客户端从伪DHCP服务器获得了错误的地址。测试如图3-3所示。

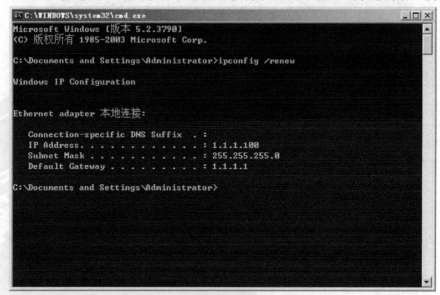

图3-3　获取伪DHCP服务器的IP地址

（4）配置DHCP监听

1）在SW1交换机上配置DHCP监听。

```
SW1(config)#ip dhcp snooping                                  #启用DHCP snooping功能
SW1(config)#interface fastEthernet 0/1
SW1(config-FastEthernet 0/1)#ip dhcp snooping trust           #配置F0/1为trust端口
SW1(config-FastEthernet 0/1)#exit
```

```
SW1(config)#interface fastEthernet 0/24
SW1(config-FastEthernet 0/24)#ip dhcp snooping trust          #配置F0/24为trust端口
SW1(config-FastEthernet 0/24)#exit
```
2）在SW2交换机上配置DHCP监听。
```
SW2(config)#ip dhcp snooping                                  #启用DHCP snooping功能
SW2(config)#interface fastEthernet 0/24
SW2(config-FastEthernet 0/24)#ip dhcp snooping trust          #配置F0/24为trust端口
SW2(config-FastEthernet 0/24)#exit
```
（5）验证测试

将客户端之前获得的错误IP地址释放（ipconfig/release），再使用ipconfig/renew重新获取地址，可以看到客户端获取了正确的IP地址。测试如图3-4所示。

图3-4　获取合法服务器分配的IP地址

由于配置了DHCP监听，并且伪DHCP服务器连接的端口为非信任端口，所以交换机丢弃了伪DHCP服务器发送的响应报文。

通过DHCP监听的学习，主要掌握DHCP监听的启用、信任端口、DHCP监听数据库信息的绑定、DHCP监听的MAC验证等配置与管理工作。

DHCP监听能够通过过滤网络中接入的伪DHCP发送的DHCP报文增强网络的安全性。DHCP监听还可以检查DHCP客户端发送的DHCP报文的合法性，防止DHCP DoS攻击。

3.3　ARP检查

问题描述

在学校网络中，网络管理员经常收到有教师反映无法访问互联网的情况，经过故障排查后，发现客户端PC缓存的网关的ARP绑定条目是错误的，可以判断网络中出现了ARP攻击。所以需要在静态IP地址环境下使用ARP检查，在动态地址分配环境中使用DAI（Dynamic ARP Inspection，动态ARP检测），以防止网络中的ARP攻击。

问题分析

ARP检查要依赖于端口安全特性，也就是说交换机的端口具有防范ARP欺骗的功能，首先要开启端口安全功能。DAI需要依赖于DHCP监听特性，所以在配置DAI之前要启用交换机的DHCP监听功能，并正确地配置端口的信任状态。

ARP技术主要涉及ARP检查配置和动态ARP检测配置。

1. ARP检查配置

步骤1 启用ARP检查功能。
switch(config)#**port-security arp-check**
步骤2 启用端口安全功能。
switch(config-if)#**switchport port-security**
步骤3 配置安全地址绑定。
switch(config-if)#**switchport port-security mac-address** *mac-address* **ip-address** *ip-address*

2. DAI配置

步骤1 开启DHCP监听功能。
switch(config)#**ip dhcp snooping**
步骤2 设置DHCP监听信任端口。需要将DHCP服务器的端口与交换机上行链路端口设置为信任端口。
switch(config-if)#**ip dhcp snooping trust**
步骤3 启用交换机的DAI功能。
switch(config)#**ip arp inspection**
步骤4 对特定VLAN启用DAI。
switch(config)#**ip arp inspection vlan** *vlan-range*
步骤5 设置DAI信任端口。
switch(config-if)#**ip arp inspection trust**

任 务 单

1	配置ARP检查
2	配置动态ARP检测

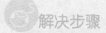解决步骤

根据任务单的安排完成任务。

任务1：配置ARP检查

任务实施

1. 任务描述及网络拓扑设计

根据图3-5所示，正确配置PC、攻击机、路由器的IP地址，保证各设备之间的互联互通。在交换机连接攻击者PC的端口上启用ARP检查功能，防止ARP欺骗攻击。

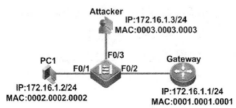

图3-5　配置ARP检查

2. 网络设备配置

```
switch#configure terminal
switch(config)#port-security arp-check            #启用ARP检查功能
switch(config)#interface fastEthernet 0/3
switch(config-if)#switchport port-security        #配置端口安全
switch(config-if)#switchport port-security mac-address 0003.0003.0003 ip-address 172.16.1.3
    #为ARP检查配置安全地址绑定
switch(config-if)#exit
```

任务2：配置动态ARP检测

任务实施

1. 任务描述及网络拓扑设计

设置服务器的地址为10.1.20.2，服务器中分配的地址池为10.1.10.0/24。SW2交换机所连接的客户机和攻击者分别设置为自动获取IP地址，分别接入交换机的F0/1和F0/2端口，绘制拓扑结构图，如图3-6所示。

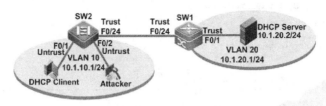

图3-6　配置动态ARP检测

2. 网络设备配置

（1）在SW1交换机配置VLAN、端口划分以及给VLAN配置IP地址

```
SW1(config)#vlan 10
SW1(config-vlan)#exit
SW1(config)#vlan 20
SW1(config-vlan)#exit
```

```
SW1(config)#interface fastEthernet 0/1
SW1(config-FastEthernet 0/1)#switchport access vlan 20
SW1(config-FastEthernet 0/1)#exit
SW1(config)#interface fastEthernet 0/24
SW1(config-FastEthernet 0/24)#switchport mode trunk
SW1(config-FastEthernet 0/24)#exit
SW1(config)#interface vlan 10
SW1(config-VLAN 10)#ip address 10.1.10.1 255.255.255.0
SW1(config-VLAN 10)#exit
SW1(config)#interface vlan 20
SW1(config-VLAN 20)#ip address 10.1.20.1 255.255.255.0
SW1(config-VLAN 20)#exit
```
（2）在SW2交换机配置VLAN以及端口划分
```
SW2(config)#vlan 10
SW2(config-vlan)#exit
SW2(config)#interface fastEthernet 0/24
SW2(config-FastEthernet 0/24)#switchport mode trunk
SW2(config-FastEthernet 0/24)#exit
SW2(config)#interface range fastEthernet 0/1-2
SW2(config-if-range)#switchport access vlan 10
```
（3）在SW1交换机上配置DHCP Rely
```
SW1(config)#service dhcp                         #启用DHCP服务
SW1(config)#interface vlan 10                    #进入VLAN接口模式
SW1(config-VLAN 10)# ip help-address 10.1.20.2   #配置DHCP中继
```
（4）在SW1交换机上配置DHCP监听
```
SW1(config)#ip dhcp snooping                              #启用DHCP snooping功能
SW1(config)#interface fastEthernet 0/1
SW1(config-FastEthernet 0/1)#ip dhcp snooping trust       #定义为DHCP监听信任端口
SW1(config-FastEthernet 0/1)#exit
SW1(config)#interface fastEthernet 0/24
SW1(config-FastEthernet 0/24)#ip dhcp snooping trust      #定义为DHCP监听信任端口
SW1(config-FastEthernet 0/24)#exit
SW1(config)#
```
（5）在SW2交换机上配置DHCP监听
```
SW2(config)#ip dhcp snooping                              #启用DHCP snooping功能
SW2(config)#interface fastEthernet 0/24
SW2(config-FastEthernet 0/24)#ip dhcp snooping trust      #定义为DHCP监听信任端口
SW2(config-FastEthernet 0/24)#exit
SW2(config)#
```
（6）在SW2交换机上配置DAI
```
SW2(config)#ip arp inspection                             #配置动态ARP检查
SW2(config)#ip arp inspection vlan 10                     #在vlan10上启用DAI
SW2(config)#interface fastEthernet 0/24
SW2(config-FastEthernet 0/24)#ip arp inspection trust     #定义为ARP检查信任端口
SW2(config-FastEthernet 0/24)#exit
```
（7）验证测试

在Attacker攻击机上安装ARP欺骗工具软件进行测试。启用ARP攻击以后，由于启用了ARP检查功能，当交换机端口收到非法ARP报文后，会将其丢失。这时在客户机上查看

ARP缓存表，发现ARP表中的条目是正确的，测试如图3-7所示。

图3-7 测试动态ARP检测

小结

通过ARP检查的学习，主要掌握ARP检查与动态ARP检查等配置与管理工作。

ARP检查依赖于端口安全特性，也就是说要使交换机的端口具有防范ARP欺骗攻击的功能，首先要启用端口安全特性。动态ARP检查要启用交换机的DHCP监听功能，并正确地配置端口的信任状态。

3.4 ACL应用

问题描述

该单位为了使校园网络用户有一个安全的操作环境，防止一些病毒的入侵和黑客的攻击；以及为了更好地规范网络用户的行为，使用IP访问控制列表对数据流进行过滤，限制网络中的通信数据类型及网络的使用者或使用设备。

问题分析

访问控制列表最直接的功能是包过滤。通过接入控制列表可以在路由器、三层交换机上进行网络安全属性配置，可以实现对进入路由器、三层交换机的输入数据流进行过滤。访问控制列表的应用方法是入栈（in）应用和出栈（out）应用。

配置ACL应用时，主要涉及编号的标准ACL配置、命名的标准ACL配置、编号的扩展ACL配置和命名的扩展ACL配置。

1. 编号的标准ACL

步骤1 定义编号的标准访问控制列表。list-number是规则序号，序号范围是1～99或者1300～1399。source代表源地址，source-wildcard表示需要检测的源IP地址的反向子网掩码。

router(config)#**access-list** *list-number* {**permit** | **deny**} {*source source-wildcard*}

步骤2 应用ACL。参数{ in | out }表示在此接口上是对哪个方向的数据进行过滤，in表示对进入接口的数据进行过滤，out表示对发出接口的数据进行过滤。

router(config)#interface *interface-id*
router(config-if)#ip access-group *list-number* { in | out }

2. 命名的标准ACL

步骤1 定义命名的标准访问控制列表。name是ACL的名称。

router(config)#**ip access-list standard** *name*
router(config-std-nacl)#{**permit** | **deny**} {*source source-wildcard*}

步骤2 应用ACL。

router(config)#interface *interface-id*
router(config-if)#ip access-group *name* { in | out }

3. 编号的扩展ACL

步骤1 定义编号的扩展访问控制列表。list-number是规则序号，序号范围是100～199或者2000～2699。source代表源地址，source-wildcard表示需要检测的源IP地址的反向子网掩码。destination代表目的地址，destination-wildcard表示需要检测的目的IP地址的反向子网掩码。host表示一种精确的匹配，只指定一个特定主机，其屏蔽码为0.0.0.0。any是0.0.0.0/255.255.255.255的简写，即任何地址都匹配语句。operator表示端口操作符，port是端口号。

router(config)#**access-list** *list-number* {**permit** | **deny**} *protocol* {*source source-wildcard* | **host** *source* | any}
[*operator port*] {*destination destination-wildcard* | **host** *destination* | any}[*operator port*]

步骤2 应用ACL。

router(config)#interface *interface-id*
router(config-if)#ip access-group *list-number* { in | out }

4. 命名的扩展ACL

步骤1 定义命名的扩展访问控制列表。

router(config)#**ip access-list extended** *name*
router(config-std-nacl)#{**permit** | **deny**} *protocol* {*source source-wildcard* | **host** *source* | any}
[*operator port*] {*destination destination-wildcard* | **host** *destination* | any}[*operator port*]

步骤2 应用ACL。

router(config-if)#ip access-group *name* { in | out }

5. 基于时间的ACL

步骤1 创建时间段。time-range-name表示时间段的名称。

router(config)#**time-range** *time-range-name*

步骤2 配置绝对时间段。start time date表示时间段的起始时间，end time date表示时间段的结束时间，在配置时间段时，可以只配置起始时间，或者只配置结束时间。

router(config-time-range)#**absolute** {**start** time date[**end** time date] | **end** time date}

步骤3 配置周期时间段。day-of-the-week表示一个星期内的一天或者几天，hh:mm表示时间，weekdays表示周一至周五，weekend表示周六到周日，daily表示一周中的某一天。

router(config-time-range)#**periodic** day-of-the-week hh:mm to [day-of-the-week] hh:mm

router(config-time-range)#**periodic** {**weekdays** | **weekend** | **daily**} hh:mm to hh:mm

步骤4 应用时间段。

配置完时间段后，在ACL规则中使用time-range参数引用时间段后才会生效，但是只有配置了time-range的规则才会在指定的时间段内生效，其他未引用时间段的规则将不受影响。

任 务 单

1	配置标准访问控制列表
2	配置扩展访问控制列表
3	配置基于时间的访问控制列表

解决步骤

根据任务单的安排完成任务。

任务实施

1. 任务描述及网络拓扑设计

要求192.168.1.0网段可以对192.168.4.0网段进行访问，但是192.168.2.0网段不可以对192.168.4.0网段进行访问。绘制拓扑结构图，如图3-8所示。

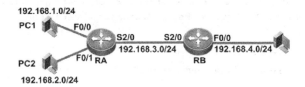

图3-8 配置标准访问控制列表

2. 网络设备配置

（1）配置路由器各接口的IP地址

1）在RA路由器上配置IP地址。

RA(config)#interface fastEthernet 0/0
RA(config-if-FastEthernet 0/0)#ip address 192.168.1.1 255.255.255.0
RA(config-if-FastEthernet 0/0)#exit
RA(config)#interface fastEthernet 0/1
RA(config-if-FastEthernet 0/1)#ip address 192.168.2.1 255.255.255.0
RA(config-if-FastEthernet 0/1)#exit
RA(config)#interface serial 2/0
RA(config-if-Serial 2/0)#ip address 192.168.3.1 255.255.255.0

2）在RB路由器上配置IP地址。
RB(config)#interface fastEthernet 0/0
RB(config-if-FastEthernet 0/0)#ip address 192.168.4.1 255.255.255.0
RB(config-if-FastEthernet 0/0)#exit
RB(config)#interface serial 2/0
RB(config-if-Serial 2/0)#ip address 192.168.3.2 255.255.255.0

（2）在RA与RB路由器上配置路由
1）在RA路由器上配置路由。
RA(config)#ip route 0.0.0.0 0.0.0.0 192.168.3.2   #配置访问RB默认路由
2）在RB路由器上配置路由。
RB(config)#ip route 0.0.0.0 0.0.0.0 192.168.3.1   #配置访问RA默认路由

（3）配置标准访问控制列表
RB(config)#access-list 10 permit 192.168.1.0 0.0.0.255
            #配置访问列表，允许192.168.1.0网段数据包通过
RB(config)#access-list 10 deny 192.168.2.0 0.0.0.255
            #配置访问列表，不允许192.168.2.0网段数据包通过

（4）应用ACL
RB(config)#interface fastEthernet 0/0
RB(config-if-fastEthernet 0/0)#ip access-group 10 out
            #接口应用访问控制列表

（5）连通性测试
1）在RA路由器上以源地址192.168.1.1测试ping 192.168.4.1
RA#ping 192.168.4.1 source 192.168.1.1
Sending 5, 100-byte ICMP Echoes to 192.168.4.1, timeout is 2 seconds:
  < press Ctrl+C to break >
!!!!!
Success rate is 100 percent (5/5), round-trip min/avg/max = 30/30/30 ms
测试结果说明，192.168.1.0网段可以对192.168.4.0网段进行访问。
2）在RA路由器上以源地址192.168.2.1测试ping 192.168.4.1
RA#ping 192.168.4.1 source 192.168.2.1
Sending 5, 100-byte ICMP Echoes to 192.168.4.1, timeout is 2 seconds:
  < press Ctrl+C to break >
.
Success rate is 0 percent (0/1)
测试结果说明，192.168.2.0网段不可以对192.168.4.0网段进行访问。

任务2：配置扩展访问控制列表

任务实施

1. 任务描述及网络拓扑设计

在学校园网络中，192.168.4.0网段为服务器群用户。要求192.168.1.0网段可以对FTP服务进行访问，但是不可以对WWW服务进行访问，192.168.2.0网段只可以对WWW服务和FTP服务进行访问，不可以访问服务器群的其他服务。绘制拓扑结构图，如图3-9所示。

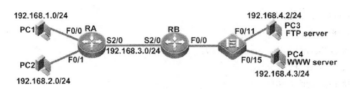

图3-9　配置扩展访问控制列表

2．网络设备配置

（1）配置路由器各接口的IP地址

1）在RA路由器上配置IP地址。

RA(config)#interface fastEthernet 0/0
RA(config-if-FastEthernet 0/0)#ip address 192.168.1.1 255.255.255.0
RA(config-if-FastEthernet 0/0)#exit
RA(config)#interface fastEthernet 0/1
RA(config-if-FastEthernet 0/1)#ip address 192.168.2.1 255.255.255.0
RA(config-if-FastEthernet 0/1)#exit
RA(config)#interface serial 2/0
RA(config-if-Serial 2/0)#ip address 192.168.3.1 255.255.255.0
RA(config-if-Serial 2/0)#exit

2）在RB路由器上配置IP地址。

RB(config)#interface fastEthernet 0/0
RB(config-if-FastEthernet 0/0)#ip address 192.168.4.1 255.255.255.0
RB(config-if-FastEthernet 0/0)#exit
RB(config)#interface serial 2/0
RB(config-if-Serial 2/0)#ip address 192.168.3.2 255.255.255.0
RB(config-if-Serial 2/0)#exit

（2）在RA与RB路由器上配置路由

1）在RA路由器上配置路由。

RA(config)#ip route 0.0.0.0 0.0.0.0 192.168.3.2

2）在RB路由器上配置路由。

RB(config)#ip route 0.0.0.0 0.0.0.0 192.168.3.1

（3）配置扩展访问控制列表

RB(config)#ip access-list extended 100
RB(config-ext-nacl)#permit tcp 192.168.1.0 0.0.0.255 host 192.168.4.2 eq ftp
RB(config-ext-nacl)#permit tcp 192.168.1.0 0.0.0.255 host 192.168.4.2 eq ftp-data
RB(config-ext-nacl)#deny tcp 192.168.1.0 0.0.0.255 host 192.168.4.3 eq www
　　　#配置访问列表，192.168.1.0网段允许访问FTP服务，不可以访问WWW服务
RB(config-ext-nacl)#permit tcp 192.168.2.0 0.0.0.255 host 192.168.4.2 eq ftp
RB(config-ext-nacl)#permit tcp 192.168.2.0 0.0.0.255 host 192.168.4.2 eq ftp-data
RB(config-ext-nacl)#permit tcp 192.168.2.0 0.0.0.255 host 192.168.4.3 eq www
RB(config-ext-nacl)#deny ip 192.168.2.0 0.0.0.255 192.168.4.0 0.0.0.255
　　　#配置访问列表，192.168.2.0网段允许访问FTP和WWW服务，不能访问其他服务
RB(config-ext-nacl)#permit ip any any
　　　#配置访问列表，允许不匹配上述规则的所有数据包通过

（4）应用ACL

RB(config)#interface fastEthernet 0/0
RB(config-if-FastEthernet 0/0)#ip access-group 100 in　　　　　　　　#接口应用访问控制列表

第3章 网络安全技术

（5）验证测试

在PC3主机上安装好FTP服务器，在PC4主机上安装好WWW服务器，然后在PC1和PC2主机上分别访问FTP服务器和WWW服务器，观察效果。

1）首先在PC1主机上配置正确的IP地址，如图3-10所示，然后访问FTP服务器和WWW服务器，效果分别如图3-11和图3-12所示。

图3-10　PC1的IP地址设置

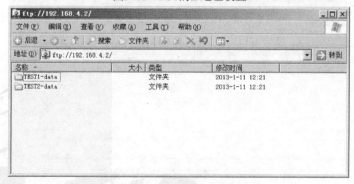

图3-11　PC1访问PC3的FTP服务

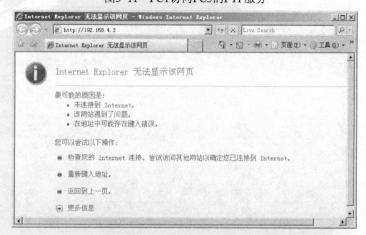

图3-12　PC1访问PC4的WWW服务

从测试结果可以看出，PC1可以访问PC3服务器的FTP服务，不可以访问PC4服务器的WWW服务。

2）首先在PC2主机上配置正确的IP地址，如图3-13所示，然后访问FTP服务器和WWW服务器，效果分别如图3-14和图3-15所示。

图3-13　PC1的IP地址设置

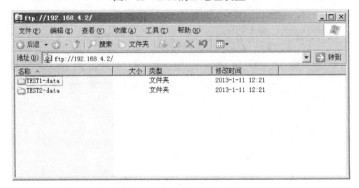

图3-14　PC2访问PC3的FTP服务

图3-15　PC2访问PC4的WWW服务

第3章 网络安全技术

从测试结果可以看出,PC2可以访问PC3服务器的FTP服务和PC4服务器的WWW服务。

任务3:配置基于时间的访问控制列表

任务实施

1. 任务描述及网络拓扑设计

在该校园网络中,要求192.168.2.0网段只有在上班时间(周一至周五8:00~17:00)才可以对WWW服务和FTP服务进行访问,其他访问不受影响。绘制拓扑结构图,如图3-16所示。

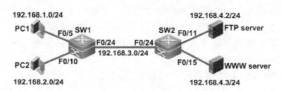

图3-16 配置基于时间的访问控制列表

2. 网络设备配置

(1)配置三层交换机各接口的IP地址

1)在SW1交换机上配置IP地址。

```
SW1(config)#interface fastEthernet 0/24
SW1(config-FastEthernet 0/24)#no switchport
SW1(config-FastEthernet 0/24)#ip address 192.168.3.1 255.255.255.0
SW1(config-FastEthernet 0/24)#exit
SW1(config)#interface fastEthernet 0/5
SW1(config-FastEthernet 0/5)#no switchport
SW1(config-FastEthernet 0/5)#ip address 192.168.1.1 255.255.255.0
SW1(config-FastEthernet 0/5)#exit
SW1(config)#interface fastEthernet 0/10
SW1(config-FastEthernet 0/10)#no switchport
SW1(config-FastEthernet 0/10)#ip address 192.168.2.1 255.255.255.0
SW1(config-FastEthernet 0/10)#exit
```

2)在SW2交换机上配置IP地址。

```
SW2(config)#interface fastEthernet 0/24
SW2(config-FastEthernet 0/24)#no switchport
SW2(config-FastEthernet 0/24)#ip address 192.168.3.2 255.255.255.0
SW2(config-FastEthernet 0/24)#exit
SW2(config)#interface vlan 1
SW2(config-VLAN 1)#ip address 192.168.4.1 255.255.255.0
SW2(config-VLAN 1)#exit
SW2(config)#exit
```

(2)在SW1与SW2三层交换机上配置路由

1)在SW1路由器上配置静态路由。

```
SW1(config)#ip route 0.0.0.0 0.0.0.0 192.168.3.2
```

2)在SW2路由器上配置静态路由。

```
SW2(config)#ip route 0.0.0.0 0.0.0.0 192.168.3.1
```

(3)配置时间段

```
SW2(config)#time-range week                    #创建时间访问列表
```

SW2(config-time-range)#periodic weekdays 08:00 to 17:00　　#定义周期时间
SW2(config-time-range)#exit

（4）配置ACL

SW2(config)#ip access-list extended aaa
SW2(config-ext-nacl)#permit tcp 192.168.2.0 0.0.0.255 host 192.168.4.2 eq ftp time-range week
SW2(config-ext-nacl)#permit tcp 192.168.2.0 0.0.0.255 host 192.168.4.2 eq ftp-data time-range week
SW2(config-ext-nacl)#permit tcp 192.168.2.0 0.0.0.255 host 192.168.4.3 eq www time-range week
#上面3条命令都是创建访问列表，并应用时间限制。表示在应用时间内，允许192.168.2.0网段访问FTP服务和WWW服务

SW2(config-ext-nacl)#deny tcp 192.168.2.0 0.0.0.255 host 192.168.4.2
SW2(config-ext-nacl)#deny tcp 192.168.2.0 0.0.0.255 host 192.168.4.3
#上面2条命令表示在应用时间外，不允许192.168.2.0网段访问FTP服务和WWW服务
SW2(config-ext-nacl)#permit ip any any
#配置访问列表，允许不匹配上述规则的所有数据包通过
SW2(config-ext-nacl)#exit

（5）应用ACL

SW2(config)#interface range fastEthernet 0/24
SW2(config-if-range)#ip access-group aaa in
SW2(config-if-range)#exit

（6）验证测试

在上班时间，PC1和PC2都可以访问服务器的FTP服务和WWW服务；在下班时间，PC1可以访问服务器的FTP服务和WWW服务，PC2不可以访问服务器的FTP服务和WWW服务。

小结

通过ACL应用的学习，主要掌握标准访问控制列表、扩展访问控制列表和基于时间的访问控制列表等配置与管理工作。

访问控制列表的主要动作为允许（permit）和拒绝（deny）。主要的应用方法是入栈（in）应用和出栈（out）应用。

访问控制列表最直接的功能是包过滤。通过接入控制列表可以在路由器、三层交换机上进行网络安全属性配置，可以实现对进入路由器、三层交换机的输入数据流进行过滤。

3.5 网络地址转换技术

问题描述

电信服务提供商为主校区提供的全局IP为68.1.1.2～68.1.1.8/28，为分校区提供的全局IP为72.1.1.2/30；教育网服务提供商为主校区提供的全局IP为210.21.1.0/24网段。使用网络地址转换技术，将私有地址转换为合法的全局IP地址；使用动态端口NAT技术实现内部用户访问互联网资源；使用静态NAT技术，将Web服务器发布到互联网。

问题分析

配置网络地址转换时,主要涉及静态NAT的配置、动态NAT的配置和基于端口的NAT配置。

1. 静态NAT

步骤1 至少指定一个内部接口和一个外部接口。inside指定接口为NAT内部接口,outside指定接口为NAT外部接口。

router(config)#**interface** *interface-id*
router(config-if)#**ip nat** {**inside** | **outside** }

步骤2 配置静态转换条目。参数local-ip表示内部网络中的主机的本地IP地址,global-ip 表示外部主机看到的内部主机的全局唯一的IP地址,interface表示路由器本地接口,如果指定该参数,路由器将使用该接口的地址进行转换。

router(config)#**ip nat indide source static** *local-ip* {**interface** *interface* | *global-ip*}

2. 动态NAT

步骤1 至少指定一个内部接口和一个外部接口。
router(config)#**interface** *interface-id*
router(config-if)#**ip nat** {**inside** | **outside** }

步骤2 定义IP访问控制列表,以明确哪些报文将被进行NAT转换。
router(config)#**access-list** *access-list-number* {**permit** | **deny**}

步骤3 定义地址池,用于转换地址。pool-name表示地址池的名称,start-ip 表示地址池包含的范围中第一个IP地址,end-ip表示地址池包含的范围中最后一个IP地址,netmask表示地址池中的地址所属网络的子网掩码,prefix-length表示地址池中的地址所属于网络的子网掩码有多少值为1。

router(config)#**ip nat pool** *pool-name start-ip end-ip* {**netmask** *netmask* | **prefix-length** *prefix-length*}

步骤4 配置动态转换条目。将符合访问控制列表条件的内部本地地址转换到地址池中的内部全局地址。access-list-number表示引用的访问控制列表的编号,pool-name表示引用地址池的名称。interface表示路由器本地接口,如果指定该参数,路由器将使用该接口的地址进行转换。

router(config)#**ip nat inside source list** *access-list-number* {**interface** *interface* | **pool** *pool-name*}

3. 动态NAPT

步骤1 至少指定一个内部接口和一个外部接口。
router(config)#**interface** *interface-id*
router(config-if)#**ip nat** {**inside** | **outside** }

步骤2 定义IP访问控制列表,以明确哪些报文将被进行NAT转换。
router(config)#**access-list** *access-list-number* {**permit** | **deny**}

步骤3 定义地址池,用于转换地址。
router(config)#**ip nat pool** *pool-name start-ip end-ip* {**netmask** *netmask* | **prefix-length** *prefix-length*}

步骤4 配置动态转换条目。在配置NAPT转换中,必须使用overload关键字,这样路

由器才会将源端口也进行转换，以便达到地址超载的目的，如果不指定overload关键字，路由器将执行动态NAT转换。

router(config)#**ip nat inside source list** *access-list-number* {**interface** *interface* | **pool** *pool-name*} **overload**

4. 配置TCP负载均衡

步骤1 至少指定一个内部接口和一个外部接口。

router(config)#**interface** *interface-id*
router(config-if)#**ip nat** {**inside** | **outside**}

步骤2 定义IP访问控制列表，指定发送到哪个虚拟主机的请求被进行负载均衡。

router(config)#**access-list** *access-list-number* {**permit** | **deny**}

步骤3 为真实的内部主机定义地址池。当使用NAT进行TCP负载均衡时，必须配置type rotary关键字，以保证地址池中的地址被轮流使用。

router(config)#**ip nat pool** *pool-name start-ip end-ip* {**netmask** *netmask* | **prefix-length** *prefix-length*} **type rotary**

步骤4 定义访问控制列表与真实主机地址池之间的映射。

router(config)#**ip nat inside destination list** *access-list-number* **pool** *pool-name*

任 务 单

1	配置静态NAT
2	配置动态NAT
3	配置NAPT
4	配置TCP负载均衡

根据任务单的安排完成任务。

任务1：配置静态NAT

任务实施

1. 任务描述及网络拓扑设计

学校需要将192.168.100.4和192.168.100.6两台服务器对外网进行发布，使用的公共IP地址分别为68.1.1.4和68.1.1.6。考虑到包括安全在内的诸多因素，希望对外部隐藏内部网络。绘制拓扑结构图，如图3-17所示。

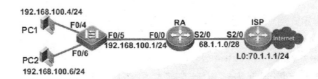

图3-17 配置静态NAT

2. 网络设备配置

（1）在路由器上配置IP地址和IP路由选择

1）在路由器RA上配置IP地址。

RA(config)#interface fastEthernet 0/0
RA(config-if-FastEthernet 0/0)#ip address 192.168.100.1 255.255.255.0
RA(config-if-FastEthernet 0/0)#exit
RA(config)#interface serial 2/0
RA(config-if-Serial 2/0)#ip address 68.1.1.2 255.255.255.240
RA(config-if-Serial 2/0)#exit

2）在路由器ISP上配置IP地址。

ISP(config)#interface serial 2/0
ISP(config-if-Serial 2/0)#ip address 68.1.1.1 255.255.255.240
ISP(config-if-Serial 2/0)#exit
ISP(config)#interface loopback 0
ISP(config-if-Loopback 0)#ip address 70.1.1.1 255.255.255.0
ISP(config-if-Loopback 0)#exit

3）在路由器RA上配置默认路由。

RA(config)#ip route 0.0.0.0 0.0.0.0 68.1.1.1   #配置访问外网默认路由

（2）配置静态NAT

RA(config)#ip nat inside source static 192.168.100.4 68.1.1.4
      #配置静态NAT，将内部服务器发布到互联网
RA(config)#ip nat inside source static 192.168.100.6 68.1.1.6
      #配置静态NAT，将内部服务器发布到互联网

（3）指定一个内部接口和一个外部接口

RA(config)#interface fastEthernet 0/0
RA(config-if-FastEthernet 0/0)#ip nat inside   #定义接口为内部接口
RA(config-if-FastEthernet 0/0)#exit
RA(config)#interface serial 2/0
RA(config-if-Serial 2/0)#ip nat outside   #定义接口为外部接口
RA(config-if-Serial 2/0)#exit

（4）验证测试

1）设置PC1的IP地址为192.168.100.4/24，在PC1上ping测试ISP上L0接口地址70.1.1.1，然后在RA路由器上观察NAT效果。

```
RA#show ip nat translations
Pro Inside global        Inside local      Outside local    Outside global
icmp68.1.1.4:512         192.168.100.4:512    70.1.1.1         70.1.1.1
```

2）设置PC2的IP地址为192.168.100.6/24，在PC2上ping测试ISP上L0接口地址70.1.1.1，然后在RA路由器上观察NAT效果。

```
RA#show ip nat translations
Pro Inside global        Inside local      Outside local    Outside global
icmp68.1.1.6:512         192.168.100.6:512    70.1.1.1         70.1.1.1
```

任务2：配置动态NAT

任务实施

1. 任务描述及网络拓扑设计

ISP提供商给该校主校区的公共IP地址的地址段是210.21.1.100～210.21.1.200/24，需要内网使用此地址段访问Internet，考虑到安全在内的诸多因素，希望对外部隐藏内部网络。将

PC1用户的IP地址设置为192.168.10.2/24，网关设置为192.168.10.1；PC2用户的IP地址设置为192.168.20.2/24，网关设置为192.168.20.1。绘制拓扑结构图，如图3-18所示。

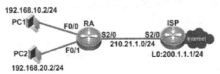

图3-18　配置动态NAT

2. 网络设备配置

（1）在路由器上配置IP地址和IP路由选择

1）在路由器RA上配置IP地址。

RA(config)#interface fastEthernet 0/0
RA(config-if-FastEthernet 0/0)#ip address 192.168.10.1 255.255.255.0
RA(config-if-FastEthernet 0/0)#exit
RA(config)#interface fastEthernet 0/1
RA(config-if-FastEthernet 0/1)#ip address 192.168.20.1 255.255.255.0
RA(config-if-FastEthernet 0/1)#exit
RA(config)#interface serial 2/0
RA(config-if-Serial 2/0)#ip address 210.21.1.1 255.255.255.0
RA(config-if-Serial 2/0)#exit

2）在路由器ISP上配置IP地址。

ISP(config)#interface serial 2/0
ISP(config-if-Serial 2/0)#ip address 210.21.1.2 255.255.255.0
ISP(config-if-Serial 2/0)#exit
ISP(config)#interface loopback 0
ISP(config-if-Loopback 0)#ip address 200.1.1.1 255.255.255.0
ISP(config-if-Loopback 0)#exit

3）在路由器RA上配置默认路由。

RA(config)#ip route 0.0.0.0 0.0.0.0 210.21.1.2   #配置访问外网默认路由

（2）定义IP访问列表

RA(config)#ip access-list standard 10   #配置允许访问外网的列表
RA(config-std-nacl)#permit any
RA(config-std-nacl)#exit

（3）配置动态NAT

RA(config)#ip nat pool aaa 210.21.1.100 210.21.1.200 prefix-length 24
　　　　#配置NAT地址池
RA(config)#ip nat inside source list 10 pool aaa overload
　　　　#配置动态NAT，允许内网访问互联网

（4）指定一个内部接口和一个外部接口

RA(config)#interface range fastEthernet 0/0-1
RA(config-if-range)#ip nat inside   #定义接口为内部接口
RA(config-if-range)#exit
RA(config)#interface serial 2/0
RA(config-if-Serial 2/0)#ip nat outside   #定义接口为外部接口
RA(config-if-Serial 2/0)#exit

（5）验证测试

第3章 网络安全技术

1) 设置PC1的IP地址为192.168.10.2/24，在PC1上ping测试ISP上L0接口地址200.1.1.1，然后在RA路由器上观察NAT效果。

RA#show ip nat translations
Pro Inside global   Inside local   Outside local   Outside global
icmp210.21.1.124:512   192.168.10.2:512   200.1.1.1   200.1.1.1

2) 设置PC2的IP地址为192.168.20.2/24，在PC2上ping测试ISP上L0接口地址200.1.1.1，然后在RA路由器上观察NAT效果。

RA#show ip nat translations
Pro Inside global   Inside local   Outside local   Outside global
icmp210.21.1.159:512   192.168.20.2:512   200.1.1.1   200.1.1.1

任务3：配置NAPT

任务实施

1. 任务描述及网络拓扑设计

由于IPv4地址不足，ISP提供商只给了学校分校区的广域网的接口地址，地址为72.1.1.2/30，需要校园网内部能够使用接口IP地址访问Internet，考虑到安全在内的诸多因素，希望对外部隐藏内部网络。绘制拓扑结构图，如图3-19所示。

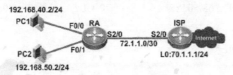

图3-19　配置NAPT

2. 网络设备配置

（1）在路由器上配置IP地址和IP路由选择

1）在路由器RA上配置IP地址。

RA(config)#interface fastEthernet 0/0
RA(config-if-FastEthernet 0/0)#ip address 192.168.40.1 255.255.255.0
RA(config-if-FastEthernet 0/0)#exit
RA(config)#interface fastEthernet 0/1
RA(config-if-FastEthernet 0/1)#ip address 192.168.50.1 255.255.255.0
RA(config-if-FastEthernet 0/1)#exit
RA(config)#interface serial 2/0
RA(config-if-Serial 2/0)#ip address 72.1.1.2 255.255.255.252

2）在路由器ISP上配置IP地址。

ISP(config)#interface serial 2/0
ISP(config-if-Serial 2/0)#ip address 72.1.1.1 255.255.255.252
ISP(config-if-Serial 2/0)#exit
ISP(config)#interface loopback 0
ISP(config-if-Loopback 0)#ip address 70.1.1.1 255.255.255.0

3）在路由器RA上配置默认路由。

RA(config)#ip route 0.0.0.0 0.0.0.0 72.1.1.1

（2）配置动态NAPT

RA(config)#ip access-list standard 10

RA(config-std-nacl)#permit any
RA(config-std-nacl)#exit
RA(config)#ip nat inside source list 10 interface serial 2/0 overload

（3）指定一个内部接口和一个外部接口

RA(config)#interface range fastEthernet 0/0-1
RA(config-if-range)#ip nat inside
RA(config-if-range)#exit
RA(config)#interface serial 2/0
RA(config-if-Serial 2/0)#ip nat outside

（4）验证测试

1）设置PC1的IP地址为192.168.40.2/24，在PC1上ping测试ISP上L0接口地址70.1.1.1，然后在RA路由器上观察NAT效果。

```
RA#show ip nat translations
Pro Inside global        Inside local       Outside local     Outside global
icmp72.1.1.2:512        192.168.40.2:512    70.1.1.1          70.1.1.1
```

2）设置PC2的IP地址为192.168.50.2/24，在PC2上ping测试ISP上L0接口地址70.1.1.1，然后在RA路由器上观察NAT效果。

```
RA#show ip nat translations
Pro Inside global        Inside local       Outside local     Outside global
icmp72.1.1.2:512        192.168.50.2:512    70.1.1.1          70.1.1.1
```

任务4：配置TCP负载均衡

任务实施

1. 任务描述及网络拓扑设计

该校内部有两台Web服务器，需要对外网进行发布并且要实现两台服务器的负载均衡，这两台服务器的地址分别为192.168.100.5/24和192.168.100.6/24，虚拟服务器的地址为210.21.1.88/24。绘制拓扑结构图，如图3-20所示。

图3-20　配置TCP负载均衡

2. 网络设备配置

（1）在路由器上配置IP地址和IP路由选择

1）在路由器RA上配置IP地址。

RA(config)#interface fastEthernet 0/0
RA(config-if-FastEthernet 0/0)#ip address 192.168.100.1 255.255.255.0
RA(config-if-FastEthernet 0/0)#exit
RA(config)#interface serial 2/0
RA(config-if-Serial 2/0)#ip address 210.21.1.1 255.255.255.0
RA(config-if-Serial 2/0)#exit

2）在路由器ISP上配置IP地址。

ISP(config)#interface serial 2/0

第3章 网络安全技术

```
ISP(config-if-Serial 2/0)#ip address 210.21.1.2 255.255.255.0
ISP(config-if-Serial 2/0)#exit
ISP(config)#interface loopback 0
ISP(config-if-Loopback 0)#ip address 200.1.1.1 255.255.255.0
```

3）在路由器RA上配置默认路由。

```
RA(config)#ip route 0.0.0.0 0.0.0.0 210.21.1.2
```

（2）为真实的主机定义一个IP地址池

```
RA(config)#ip nat pool web 192.168.100.5 192.168.100.6 netmask 255.255.255.0 type rotary
```

（3）定义一个标准访问列表

```
RA(config)#ip access-list standard 10
RA(config-std-nacl)#permit 210.21.1.88 0.0.0.255
```

（4）定义访问列表与真实主机地址池之间的映射

```
RA(config)#ip nat inside destination list 10 pool web
```

（5）指定一个内部接口和一个外部接口

```
RA(config)#interface fastEthernet 0/0
RA(config-if-FastEthernet 0/0)#ip nat inside
RA(config-if-FastEthernet 0/0)#exit
RA(config)#interface serial 2/0
RA(config-if-Serial 2/0)#ip nat outside
```

小结

通过网络地址转换的学习，主要掌握静态NAT转换、动态NAT转换、基于端口的NAT转换、TCP负载均衡等配置与管理工作。

NAT转换包括多种类型，它们之间的操作方式存在一些差别。静态NAT需要手工预先配置转换条目，并且是一对一的转换，即一个内部地址与一个外部地址唯一进行绑定。动态NAT将内部地址动态地转换为地址池中的地址，无需手工配置转换条目。NAPT不仅对IP地址信息进行转换，而且还对端口号也进行转换。在NAPT中，NAT设备通过使用一个或多个外部地址与不同的端口号来唯一地址映射一个内部地址，最大限度地减少了公有地址的使用数量。通过NAPT，企业只通过一个公有地址就可以实现内部网络中多个设备与Internet的连接。

第4章 远程接入技术

近年来,随着计算机网络技术的飞速发展,广域网得到了很大的发展。自20世纪80年代以来,ISO公布了OSI参考模型,提供了计算机网络通信协议的结构和标准层次划分,使得异种计算机的互联网络有了一个公认的协议准则;另外,一个计算机的高速发展,促进了LAN的标准化、产品化,使它成为WAN的一个可靠的基本组成部分。

随着网络需求的不断扩大,快速以太网、千兆以太网、万兆以太网的出现,使得无论是公司的网络业务,还是个人的业务都在不断地扩大,这也促使ISP不断地扩容其广域网的基础设施,不仅在地理范围上需要超越城市、省界、国界、洲界形成世界范围的计算机互联网络,而且在各种远程通信手段上有很大的变化,如除了原有的电话网外,已有分组数据交换网、数字数据网、帧中继网以及集话音、图像、数据等为一体的ISDN网、数字卫星网和无线分组数据通信网等。同时WAN在技术上也有许多突破,如互联设备的快速发展,多路复用技术和交换技术的发展,特别是ATM交换技术的成熟,为广域网解决传输带宽这个瓶颈问题展现了美好的前景。

本章介绍的远程接入技术主要涉及点对点协议PPP和帧中继技术。

1. 点对点协议PPP

PPP是为在同等单元之间传输数据包这样的简单链路设计的链路层协议。这种链路提供全双工操作,并按照顺序传递数据包。

PPP协议使用了OSI分层体系结构中的3层,分别是物理层、数据链路层和网络层。物理层用来实现点到点的连接,将IP数据报封装到串行链路;数据链路层用来建立和配置连接;网络层用来配置不同的网络,网络控制协议NCP(Network Control Protocal),支持不同的网络层协议。PPP使用它的链路控制协议LCP(Link Control Protocal)在广域网链路上协商和设置选项,使用网络控制程序组建对多种网络层协议进行封装及选项协商。LCP位于物理层之上,PPP也通过使用LCP来自动匹配链路两端之间封装格式选项。

PPP支持两种授权协议:PAP(Password Authentication Protocal,密码认证协议)和CHAP(Challenge Hand Authentication Protocal,握手认证协议)。PAP协议认证通过两次握手机制,为建立远程节点的验证提供了一个简单的方法。CHAP协议认证使用三次握手机制来启动一条链路和周期性的验证远程节点。

2. 帧中继技术

帧中继是一种用于连接计算机系统的面向分组的通信方法。它主要用在公共或专用网上的局域网互联以及广域网连接。

帧中继网络环境下的设备可以分为两大类,即数据终端设备(DTE)和数据电路终端设备(DCE)。DTE可以被理解成是网络的末端设备,通常被放置在用户区域,直接由用户所有和控制。DCE是由运营商所有的网络互连设备,主要用来提供网络的时钟和交换服务,可

第4章 远程接入技术

以通过广域网对数据进行传输。

4.1 点对点协议PPP

问题描述

常见的广域网专线技术有DDN专线、PSTN/ISDN专线、帧中继专线、X.25专线等。数据链路层提供各种专线技术的协议，主要有PPP、HDLC、X.25、Frame-relay及ATM等。

主校区与分校区使用DDN专线实现网络互联，为了保障链路安全，需要配置PPP，并采用先PAP后CHAP的验证方式，将主校区路由器设置为验证方。

问题分析

PAP协议认证通过两次握手机制，为建立远程节点的验证提供了一个简单的方法。如果是PAP认证方式，在PPP链路建立后，被验证方此时需要向验证方发送PAP认证的请求报文，该请求报文中携带了用户名和密码，当验证方收到该认证请求报文后，则会根据报文中的实际内容查找本地数据库，如果本地数据库中有与用户名和密码一致的选项，则向对方返回一个请求响应，告诉对方验证通过。反之，如果用户名和密码不符，则向对方返回不通过的响应报文。如果对方都配置为验证方，则需要双方的两个单项验证过程都完成后，方可进入网络层协议阶段。

CHAP协议认证使用三次握手机制来启动一条链路和周期性的验证远程节点。如果是CHAP验证方式，在PPP链路建立后，由验证方向被验证方发送一段随机的报文，并加上自己的主机名，通常这个过程叫作挑战。当被验证方收到验证方的验证请求时，从中提取出验证方所发送过来的主机名，然后根据该主机名，在被验证方设备的后台数据库中去查找相同的用户名的记录，当查找到同该用户名所对应的密钥后，再根据这个密钥、报文ID和验证方发送的随机报文用MD5加密算法生成应答，随后将应答和自己的主机名送回。同样验证方收到被验证方发送的回应后，提取被验证方的用户名，然后去查找本地的数据库，当找到与被验证方一致的用户名后，根据该用户名所对应的密钥、保留报文ID和随机报文用MD5加密算法生成结果，和刚刚被验证方所返回的应答进行比较，相同则返回配置确认，否则返回配置否认。

配置PPP协议时，主要涉及PAP认证、CHAP认证、PAP CHAP认证和CHAP PAP认证。

1. PPP协议封装

步骤1 进入接口模式配置时钟频率。bps表示时钟速率的具体值。
router(config-if)#**clock rate** *bps*

步骤2 PPP协议封装。
router(config-if)#**encapsulation ppp**

2. PAP认证配置

步骤1 服务器端建立本地密码数据库。name为用户名，password为用户密码，0|7为

密码的加密类型，0表示无加密，7表示简单加密，encrypted-password为密码文本。

router(config)#**username** *name* {**nopassword** | **password** {*password* |[**0**|**7**] *encrypted-password*} }

步骤2 服务器端要求进行PAP认证。pap表示在接口上启用PAP认证，chap表示在接口上启用CHAP认证，pap chap是同时启用PAP和CHAP认证，先执行PAP认证再执行CHAP认证，chap pap是同时启用PAP和CHAP认证，先执行CHAP认证再执行PAP认证。

router(config-if)#**ppp authentication** {**pap** | **chap** | **pap chap** | **chap pap**}

步骤3 PAP认证客户端配置，客户端将用户名和密码发送到对端。username是在PAP身份认证中发送的用户名，encryption-type是PAP身份认证中发送的密码类型，password是PAP身份认证中发送的密码。

router(config-if)#**ppp pap sent-username** *username*[**password** *encryption-type password*]

3. CHAP认证配置

步骤1 建立本地密码数据库。

router(config-if)#**username** *name* {**nopassword** | **password** {*password* |[**0**|**7**] *encrypted-password*} }

步骤2 要求进行CHAP认证。

router(config-if)#**ppp authentication** {**pap** | **chap** | **pap chap** | **chap pap**}

任 务 单

1	配置PAP认证
2	配置CHAP认证
3	配置CHAP PAP认证
4	配置双向PAP CHAP认证

解决步骤

根据任务单的安排完成任务。

任务1. 配置PAP认证

任务实施

1. 任务m描述及网络拓扑设计

在RA与RB所连接的串行链路上封装PPP协议，并采用PAP认证方式，将RA路由器设置为验证方。绘制拓扑结构图，如图4-1所示。

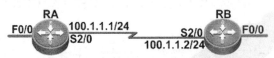

图4-1 配置PAP认证

2. 网络设备配置

（1）路由器基本配置

1）在RA路由器上配置IP地址。

RA(config)#interface serial 2/0

第4章 远程接入技术

RA(config-if-Serial 2/0)#ip address 100.1.1.1 255.255.255.0
2）在RB路由器上配置IP地址。
RB(config)#interface serial 2/0
RB(config-if-Serial 2/0)#ip address 100.1.1.2 255.255.255.0
（2）配置PAP认证
1）在RA路由器上配置PAP认证。

RA(config)#username RB password 0 123456　　　　　　　　#创建验证数据库
RA(config)#interface serial 2/0
RA(config-if-Serial 2/0)#encapsulation ppp　　　　　　　　#封装PPP协议
RA(config-if-Serial 2/0)#ppp authentication pap　　　　　　#启用PAP认证

2）在RB路由器上配置PAP认证。

RB(config)#interface serial 2/0
RB(config-if-Serial 2/0)#encapsulation ppp　　　　　　　　#封装PPP协议
RB(config-if-Serial 2/0)#ppp pap sent-username RB password 0 123456
　　　　#客户端将用户名和密码发送到服务端验证

（3）验证PAP认证

RA#show interface serial 2/0
Serial 2/0 is UP , line protocol is UP
Hardware is SIC-1HS HDLC CONTROLLER Serial
Interface address is: 100.1.1.1/24
　MTU 1500 bytes, BW 2000 Kbit
　Encapsulation protocol is PPP, loopback not set
　Keepalive interval is 10 sec , set
　Carrier delay is 2 sec
　RXload is 1 ,Txload is 1
　LCP Open
　Open: ipcp
　Queueing strategy: FIFO
　　Output queue 0/40, 0 drops;
　　Input queue 0/75, 0 drops
　　1 carrier transitions
　　V35 DTE cable
　　DCD=up DSR=up DTR=up RTS=up CTS=up
　5 minutes input rate 147 bits/sec, 0 packets/sec
　5 minutes output rate 119 bits/sec, 0 packets/sec
　　309 packets input, 6776 bytes, 0 no buffer, 0 dropped
　　Received 196 broadcasts, 0 runts, 0 giants
　　1 input errors, 0 CRC, 1 frame, 0 overrun, 0 abort
　　316 packets output, 6478 bytes, 0 underruns , 0 dropped
　　0 output errors, 0 collisions, 3 interface resets

任务2：配置CHAP认证

任务实施

1. 任务描述及网络拓扑设计

在RA与RB所连接的串行链路上封装PPP协议，并采用CHAP认证方式，将RA路由器设置为验证方。绘制拓扑结构图，如图4-2所示。

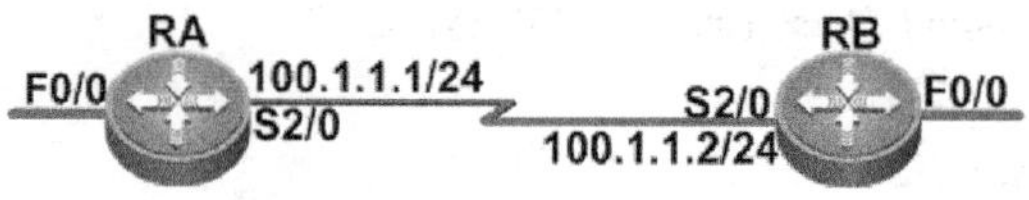

图4-2 配置CHAP认证

2. 网络设备配置

（1）路由器基本配置

1）在RA路由器上配置IP地址。

RA(config)#interface serial 2/0
RA(config-if-Serial 2/0)#ip address 100.1.1.1 255.255.255.0

2）在RB路由器上配置IP地址。

RB(config)#interface serial 2/0
RB(config-if-Serial 2/0)#ip address 100.1.1.2 255.255.255.0

（2）配置CHAP认证

1）在RA路由器上配置CHAP认证。

RA(config)#username RB password 0 123456   #创建验证数据库
RA(config)#interface serial 2/0
RA(config-if-Serial 2/0)#encapsulation ppp   #封装PPP协议
RA(config-if-Serial 2/0)#ppp authentication chap   #启用CHAP认证

2）在RB路由器上配置CHAP认证。

RB(config)#interface serial 2/0
RB(config-if-Serial 2/0)#encapsulation ppp   #封装PPP协议
RB(config-if-Serial 2/0)#ppp chap hostname RB   #指定CHAP验证的主机名
RB(config-if-Serial 2/0)#ppp chap password 0 123456   #指定CHAP验证的密码
RB(config-if-Serial 2/0)#exit

（3）验证CHAP认证

RA#show interface serial 2/0
Serial 2/0 is UP , line protocol is UP
Hardware is SIC-1HS HDLC CONTROLLER Serial
Interface address is: 100.1.1.1/24
　　MTU 1500 bytes, BW 2000 Kbit
　　Encapsulation protocol is PPP, loopback not set
　　Keepalive interval is 10 sec , set
　　Carrier delay is 2 sec
　　RXload is 1 ,Txload is 1
　　LCP Open
　　Open: ipcp
　　Queueing strategy: FIFO
　　　Output queue 0/40, 0 drops;
　　　Input queue 0/75, 0 drops
　　　1 carrier transitions
　　　V35 DTE cable
　　　DCD=up DSR=up DTR=up RTS=up CTS=up
5 minutes input rate 147 bits/sec, 0 packets/sec
5 minutes output rate 119 bits/sec, 0 packets/sec
　　309 packets input, 6776 bytes, 0 no buffer, 0 dropped
　　Received 196 broadcasts, 0 runts, 0 giants
　　1 input errors, 0 CRC, 1 frame, 0 overrun, 0 abort

      316 packets output, 6478 bytes, 0 underruns , 0 dropped
      0 output errors, 0 collisions, 3 interface resets

任务3：配置CHAP PAP认证

任务实施

1. 任务描述及网络拓扑设计

在RA与RB所连接的串行链路上封装PPP协议，并采用先CHAP后PAP验证方式，将RA路由器设置为验证方。绘制拓扑结构图，如图4-3所示。

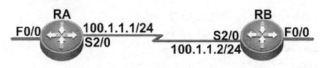

图4-3　配置CHAP PAP认证

2. 网络设备配置

（1）路由器基本配置

1）在RA路由器上配置IP地址。

RA(config)#interface serial 2/0
RA(config-if-Serial 2/0)#ip address 100.1.1.1 255.255.255.0
RA(config-if-Serial 2/0)#exit

2）在RB路由器上配置IP地址。

RB(config)#interface serial 2/0
RB(config-if-Serial 2/0)#ip address 100.1.1.2 255.255.255.0
RB(config-if-Serial 2/0)#exit

（2）配置CHAP PAP认证

1）在RA路由器上配置CHAP PAP认证。

RA(config)#username RB password 0 aaa   #创建验证数据库
RA(config)#interface serial 2/0
RA(config-if-Serial 2/0)#encapsulation ppp   #封装PPP协议
RA(config-if-Serial 2/0)#ppp authentication chap pap #启用CHAP PAP认证
RA(config-if-Serial 2/0)#exit

2）在RB路由器上配置CHAP PAP认证。

RB(config)#interface serial 2/0
RB(config-if-Serial 2/0)#encapsulation ppp
RB(config-if-Serial 2/0)#ppp chap hostname RB   #指定CHAP验证的主机名
RB(config-if-Serial 2/0)#ppp chap password 0 aaa   #指定CHAP验证的密码
RB(config-if-Serial 2/0)#ppp pap sent-username RB password 0 aaa
           #PAP认证时，客户端将用户名和密码发送到服务端验证
RB(config-if-Serial 2/0)#exit

（3）验证测试

RA#show int s 2/0
Serial 2/0 is UP   , line protocol is UP
Hardware is SIC-1HS HDLC CONTROLLER Serial
Interface address is: 100.1.1.1/24
   MTU 1500 bytes, BW 2000 Kbit

```
Encapsulation protocol is PPP, loopback not set
Keepalive interval is 10 sec , set
Carrier delay is 2 sec
RXload is 1 ,Txload is 1
LCP Open
Open: ipcp
Queueing strategy: FIFO
   Output queue 0/40, 0 drops;
   Input queue 0/75, 0 drops
   1 carrier transitions
   V35 DTE cable
   DCD=up  DSR=up  DTR=up  RTS=up  CTS=up
5 minutes input rate 147 bits/sec, 0 packets/sec
5 minutes output rate 119 bits/sec, 0 packets/sec
   309 packets input, 6776 bytes, 0 no buffer, 0 dropped
   Received 196 broadcasts, 0 runts, 0 giants
   1 input errors, 0 CRC, 1 frame, 0 overrun, 0 abort
   316 packets output, 6478 bytes, 0 underruns , 0 dropped
   0 output errors, 0 collisions, 3 interface resets
```

任务4：配置双向PAP CHAP认证

任务实施

1. 任务描述及网络拓扑设计

在RA与RB所连接的串行链路上封装PPP协议，并采用双向先PAP后CHAP验证方式。绘制拓扑结构图，如图4-4所示。

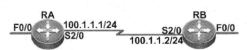

图4-4 配置PAP CHAP认证

2. 网络设备配置

（1）路由器基本配置

1）在RA路由器上配置IP地址。

RA(config)#interface serial 2/0
RA(config-if-Serial 2/0)#ip address 100.1.1.1 255.255.255.0

2）在RB路由器上配置IP地址。

RB(config)#interface serial 2/0
RB(config-if-Serial 2/0)#ip address 100.1.1.2 255.255.255.0

（2）配置PAP CHAP认证

1）在RA路由器上配置PAP CHAP认证。

RA(config)#username RB password 0 123456   #创建验证数据库
RA(config)#interface serial 2/0
RA(config-if-Serial 2/0)#encapsulation ppp   #封装PPP协议
RA(config-if-Serial 2/0)#ppp authentication pap chap   #启用PAP CHAP认证
RA(config-if-Serial 2/0)#ppp pap sent-username RA password 0 123456
           #PAP认证时，客户端将用户名和密码发送到服务端验证
RA(config-if-Serial 2/0)#ppp chap hostname RA   #指定CHAP验证的主机名

第4章 远程接入技术

RA(config-if-Serial 2/0)#ppp chap password 0 123456　　　　#指定CHAP验证的密码

2）在RB路由器上配置PAP CHAP认证。

RB(config)#username RA password 0 123456　　　　#创建验证数据库
RB(config)#interface serial 2/0
RB(config-if-Serial 2/0)#encapsulation ppp　　　　#封装PPP协议
RB(config-if-Serial 2/0)#ppp authentication pap chap　　　　#启用PAP CHAP认证
RB(config-if-Serial 2/0)#ppp pap sent-username RB password 0 123456
　　　　　#PAP认证时，客户端将用户名和密码发送到服务端验证
RB(config-if-Serial 2/0)#ppp chap hostname RB　　　　#指定CHAP验证的主机名
RB(config-if-Serial 2/0)#ppp chap password 0 123456
RB(config-if-Serial 2/0)#exit

（3）验证PAP CHAP认证

RA#show interface serial 2/0
Serial 2/0 is UP , line protocol is UP
Hardware is SIC-1HS HDLC CONTROLLER Serial
Interface address is: 100.1.1.1/24
　MTU 1500 bytes, BW 2000 Kbit
　Encapsulation protocol is PPP, loopback not set
　Keepalive interval is 10 sec , set
　Carrier delay is 2 sec
　RXload is 1 ,Txload is 1
　LCP Open
　Open: ipcp
　Queueing strategy: FIFO
　　Output queue 0/40, 0 drops;
　　Input queue 0/75, 0 drops
　　1 carrier transitions
　　V35 DTE cable
　　DCD=up DSR=up DTR=up RTS=up CTS=up
5 minutes input rate 147 bits/sec, 0 packets/sec
5 minutes output rate 119 bits/sec, 0 packets/sec
　　309 packets input, 6776 bytes, 0 no buffer, 0 dropped
　　Received 196 broadcasts, 0 runts, 0 giants
　　1 input errors, 0 CRC, 1 frame, 0 overrun, 0 abort
　　316 packets output, 6478 bytes, 0 underruns , 0 dropped
　　0 output errors, 0 collisions, 3 interface resets

小结

通过PPP协议认证的学习，主要掌握PAP认证、CHAP认证、PAP CHAP认证和CHAP PAP认证等配置与管理工作。

4.2 帧中继技术

问题描述

主校区与分校区使用城域网专用链路为主链路来实现互联，现需要使用帧中继服务实现

建立高性能的虚拟广域网连接。

问题分析

帧中继是一种用于连接计算机系统的面向分组的通信方法。它主要用在公共或专用网上的局域网互联以及广域网连接。

帧中继网络环境下的设备可以分为两大类，即数据终端设备（DTE）和数据电路终端设备（DCE）。DTE可以被理解成是网络的末端设备，通常被放置在用户区域，直接由用户所有和控制。DCE是运营商所有的网络互联设备，主要用来提供网络的时钟和交换服务，可以通过广域网对数据进行传输。通常，DCE设备主要是包交换机。

帧中继技术的配置主要包括帧中继基本配置和帧中继交换机配置。

1. 帧中继基本配置

步骤1 配置接口封装协议。系统默认封装格式是cisco格式，如果没有特殊的使用场合，则封装ietf类型。

router(config-if)#**encapsulation frame-relay [ietf]**

步骤2 配置静态映射。静态地址映射反映远端设备的IP地址和本地DLCI的对应关系。在对端设备不支持逆向ARP时，本地端必须配置静态地址映射才能通信。

router(config-if)#**frame-relay map ip** *ip-address dlci* [**broadcast** | **ietf** | **cisco**]

步骤3 配置本地管理接口，这是可选配置。系统默认是Q933A，一般局方提供ANSI类型。

router(config-if)#**frame-relay lmi-type** {**q933a** | **ansi** | **cisco**}

2. 帧中继子接口配置

步骤1 创建子接口。

router(config)#**interface serial***slot-number/interface.sub-port* [**point-to-point** | **multipoint**]

步骤2 配置帧中继子接口的DLCI号。

router(config-subif)#**frame-relay interface-dlci** *dlci*

步骤3 配置帧中继子接口PVC和建立地址映射。

router(config-subif)#**frame-relay map ip** *ip-address dlci* [*option*]

3. 帧中继交换机配置

步骤1 设定交换机交换功能命令。

router(config)#**frame-relay switching**

步骤2 设定接口为DCE类型。

router(config-if)#**frame-relay intf-type** {**dte** | **dce**}

步骤3 配置帧中继交换路由。

router(config-if)#**frame-relay route** *in-dlci* **interface** *serialnumber out-dlci*

任 务 单

1	帧中继基本配置
2	帧中继交换机配置

第4章 远程接入技术

根据任务单的安排完成任务。

任务1：帧中继基本配置

任务实施

1. 任务描述及网络拓扑设计

在RA与RB通过公用帧中继网络互联局域网，在这种方式下，路由器只能作为用户设备工作在帧中继的DTE方式。设定路由器RA的DLCI号为16，路由器RB的DLCI号为17。绘制拓扑结构图，如图4-5所示。

图4-5 帧中继基本配置

2. 网络设备配置

（1）路由器基本配置

1）在RA路由器上配置IP地址。

RA(config)#interface serial 2/0
RA(config-if-Serial 2/0)#ip address 100.1.1.1 255.255.255.0
RA(config-if-Serial 2/0)#exit
RA(config)#interface fastethernet 0/0
RA(config-if-fastethernet 0/0)#ip address 192.168.1.1 255.255.255.0
RA(config-if-fastethernet 0/0)#exit

2）在RB路由器上配置IP地址。

RB(config)#interface serial 2/0
RB(config-if-Serial 2/0)#ip address 100.1.1.2 255.255.255.0
RB(config-if-Serial 2/0)#exit
RB(config)#interface fastethernet 0/0
RB(config-if-fastethernet 0/0)#ip address 192.168.2.1 255.255.255.0
RB(config-if-fastethernet 0/0)#exit
RB(config)#

（2）配置帧中继

1）在RA路由器上配置帧中继。

RA(config)#interface serial 2/0   #进入接口模式
RA(config-if-Serial 2/0)#encapsulation frame-relay ietf   #封装帧中继协议
RA(config-if-Serial 2/0)#frame-relay lmi-type ansi   #定义LMI管理接口
RA(config-if-Serial 2/0)#frame-relay interface-dlci 16   #指定DLCI号
RA(config-if-Serial 2/0)#frame-relay map ip 100.1.1.2 16 ietf   #定义映射
RA(config-if-Serial 2/0)#exit

2）在RB路由器上配置帧中继。

RB(config)#interface serial 2/0   #进入接口模式
RB(config-if-Serial 2/0)#encapsulation frame-relay ietf   #封装帧中继协议
RB(config-if-Serial 2/0)#frame-relay lmi-type ansi   #定义LMI管理接口
RB(config-if-Serial 2/0)#frame-relay interface-dlci 17   #指定DLCI号

RB(config-if-Serial 2/0)#frame-relay map ip 100.1.1.1 17 ietf   #定义映射
RB(config-if-Serial 2/0)#exit

（3）验证帧中继配置

1）查看帧中继永久虚电路PVC信息。

RA#show frame-relay pvc

PVC Statistics for interface Serial 2/0 (Frame Relay DTE)

DLCI = 16, DLCI USAGE = LOCAL,PVC STATUS = DELETED,INTERFACE = Serial 2/0

  input pkts 183 output pkts 187 in bytes 5868
  out bytes 6350 dropped pkts 0 in FECN pkts 0
  switched pkts 0
  in BECN pkts 0 out FECN pkts 0 out BECN pkts 0
  in DE pkts 0   out DE pkts 0

DLCI = 17, DLCI USAGE = LOCAL, PVC STATUS = ACTIVE , INTERFACE = Serial 2/0

  input pkts 21 output pkts 24 in bytes 1434
  out bytes 1830 dropped pkts 0 in FECN pkts 0
  switched pkts 0
  in BECN pkts 0 out FECN pkts 0 out BECN pkts 0
  in DE pkts 0   out DE pkts 0

2）查看帧中继映射表。

RA#show frame-relay map
Serial 2/0 (up): ip 100.1.1.2
     dlci 17(0x410), dynamic,
     broadcast,IETF, status: ACTIVE
Serial 2/0 (up): ip 100.1.1.1
     dlci 16(0x400), static
     IETF, status: DELETED

3）查看帧中继本地管理信息。

RA#show frame-relay lmi

LMI Statistics for interface Serial 2/0 (Frame Relay DTE) LMI TYPE = ANSI Annex D

 Invalid Unnumbered info 0 Invalid Prot Disc 0
 Invalid dummy Call Ref 0 Invalid Msg Type 0
 Invalid Status Message 0 Invalid Lock Shift 0
 Invalid Information ID 0 Invalid Report ELE Len 0
 Invalid Report Request 0 Invalid Keepalive ELE Len 0
 Num Status Enq. Sent 179 Num Status msgs Rcvd 170
 Num Update Status Rcvd 0 Num Status Timeouts 0

任务2：帧中继交换机的配置

任务实施

1. 任务描述及网络拓扑设计

利用路由器来配置帧中继交换机，在帧中继环境下，设定路由器RA的DLCI号为16和

17,路由器RB的DLCI号为26和27,路由器RC的DLCI号为36和37。绘制拓扑结构图,如图4-6所示。

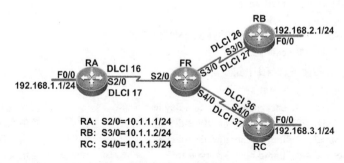

图4-6 帧中继交换机配置

2. 网络设备配置

(1)RA路由器配置

1)在RA路由器上配置IP地址。

RA(config)#interface fastEthernet 0/0
RA(config-if-FastEthernet 0/0)#ip address 192.168.1.1 255.255.255.0
RA(config-if-FastEthernet 0/0)#exit
RA(config)#interface serial 2/0
RA(config-if-Serial 2/0)#ip address 10.1.1.1 255.255.255.0
RA(config-if-Serial 2/0)#exit

2)在RA路由器上配置帧中继。

RA(config)#interface serial 2/0
RA(config-if-Serial 2/0)#encapsulation frame-relay ietf   #封装帧中继协议
RA(config-if-Serial 2/0)#frame-relay interface-dlci 16   #指定DLCI号
RA(config-if-Serial 2/0)#frame-relay interface-dlci 17   #指定DLCI号
RA(config-if-Serial 2/0)#rame-relay lmi-type ansi   #定义LMI管理接口
RA(config-if-Serial 2/0)#exit

(2)RB路由器配置

1)在RB路由器上配置IP地址。

RB(config)#interface fastEthernet 0/0
RB(config-if-FastEthernet 0/0)#ip address 192.168.2.1 255.255.255.0
RB(config-if-FastEthernet 0/0)#exit
RB(config)#interface serial 3/0
RB(config-if-Serial 3/0)#ip address 10.1.1.2 255.255.255.0
RB(config-if-Serial 3/0)#exit

2)在RB路由器上配置帧中继。

RB(config)#interface serial 3/0
RB(config-if-Serial 3/0)#encapsulation frame-relay ietf   #封装帧中继协议
RB(config-if-Serial 3/0)#frame-relay interface-dlci 26   #指定DLCI号
RB(config-if-Serial 3/0)#frame-relay interface-dlci 27   #指定DLCI号
RB(config-if-Serial 3/0)#frame-relay lmi-type ansi   #定义LMI管理接口
RB(config-if-Serial 3/0)#exit

(3)RC路由器配置

1)在RC路由器上配置IP地址。

```
RC(config)#interface fastEthernet 0/0
RC(config-if-FastEthernet 0/0)#ip address 192.168.3.1 255.255.255.0
RC(config-if-FastEthernet 0/0)#exit
RC(config)#interface serial 4/0
RC(config-if-Serial 4/0)#ip address 10.1.1.3 255.255.255.0
RC(config-if-Serial 4/0)#exit
```

2）在RC路由器上配置帧中继。

```
RC(config)#interface serial 4/0
RC(config-if-Serial 4/0)#encapsulation frame-relay ietf          #封装帧中继协议
RC(config-if-Serial 4/0)#frame-relay interface-dlci 36           #指定DLCI号
RC(config-if-Serial 4/0)#frame-relay interface-dlci 37           #指定DLCI号
RC(config-if-Serial 4/0)#frame-relay lmi-type ansi               #定义LMI管理接口
RC(config-if-Serial 4/0)#exit
```

（4）配置帧中继交换机

```
FR(config)#frame-relay switching                                 #设定交换机交换功能命令
FR(config)#interface serial 2/0
FR(config-if-Serial 2/0)#encapsulation frame-relay ietf          #封装帧中继协议
FR(config-if-Serial 2/0)#frame-relay lmi-type ansi               #定义LMI管理接口
FR(config-if-Serial 2/0)#frame-relay intf-type dce               #设定接口为dce类型
FR(config-if-Serial 2/0)#frame-relay route 26 interface serial 3/0 16   #配置帧中继交换路由
FR(config-if-Serial 2/0)#frame-relay route 37 interface serial 4/0 17   #配置帧中继交换路由
FR(config-if-Serial 2/0)#exit
FR(config)#interface serial 3/0
FR(config-if-Serial 3/0)#encapsulation frame-relay ietf          #封装帧中继协议
FR(config-if-Serial 3/0)#frame-relay lmi-type ansi               #定义LMI管理接口
FR(config-if-Serial 3/0)#frame-relay intf-type dce               #设定接口为dce类型
FR(config-if-Serial 3/0)#frame-relay route 16 interface serial 2/0 26   #配置帧中继交换路由
FR(config-if-Serial 3/0)#frame-relay route 36 interface serial 4/0 27   #配置帧中继交换路由
FR(config-if-Serial 3/0)#exit
FR(config)#interface serial 4/0
FR(config-if-Serial 4/0)#encapsulation frame-relay ietf          #封装帧中继协议
FR(config-if-Serial 4/0)#frame-relay lmi-type ansi               #设定接口为dce类型
FR(config-if-Serial 4/0)#frame-relay intf-type dce               #配置帧中继交换路由
FR(config-if-Serial 4/0)#frame-relay route 17 interface serial 2/0 37   #配置帧中继交换路由
FR(config-if-Serial 4/0)#frame-relay route 27 interface serial 3/0 36   #配置帧中继交换路由
FR(config-if-Serial 4/0)#exit
```

（5）验证测试

1）在RA设备上ping命令测试。

```
RA#ping 10.1.1.2
Sending 5, 100-byte ICMP Echoes to 10.1.1.2, timeout is 2 seconds:
   < press Ctrl+C to break >
!!!!!
Success rate is 100 percent (5/5), round-trip min/avg/max = 60/62/70 ms
RA#ping 10.1.1.3
Sending 5, 100-byte ICMP Echoes to 10.1.1.3, timeout is 2 seconds:
   < press Ctrl+C to break >
!!!!!
Success rate is 100 percent (5/5), round-trip min/avg/max = 60/62/70 ms
```

2）在RB设备上ping命令测试。

```
RB#ping 10.1.1.1
Sending 5, 100-byte ICMP Echoes to 10.1.1.1, timeout is 2 seconds:
   < press Ctrl+C to break >
!!!!!
Success rate is 100 percent (5/5), round-trip min/avg/max = 60/62/70 ms
RB#ping 10.1.1.3
Sending 5, 100-byte ICMP Echoes to 10.1.1.3, timeout is 2 seconds:
   < press Ctrl+C to break >
!!!!!
Success rate is 100 percent (5/5), round-trip min/avg/max = 60/62/70 ms
```

3）在RC设备上ping命令测试。

```
RC#ping 10.1.1.2
Sending 5, 100-byte ICMP Echoes to 10.1.1.2, timeout is 2 seconds:
   < press Ctrl+C to break >
!!!!!
Success rate is 100 percent (5/5), round-trip min/avg/max = 60/62/70 ms
RC#ping 10.1.1.1
Sending 5, 100-byte ICMP Echoes to 10.1.1.1, timeout is 2 seconds:
   < press Ctrl+C to break >
!!!!!
Success rate is 100 percent (5/5), round-trip min/avg/max = 60/62/70 ms
```

4）查看帧中继交换路由信息。

```
FR#show frame-relay route
```

Input Intf	Input Dlci	Output Intf	Output Dlci	Status
Serial 2/0	26	Serial 3/0	16	ACTIVE
Serial 2/0	37	Serial 4/0	17	ACTIVE
Serial 3/0	16	Serial 2/0	26	ACTIVE
Serial 3/0	36	Serial 4/0	27	ACTIVE
Serial 4/0	17	Serial 2/0	37	ACTIVE
Serial 4/0	27	Serial 3/0	36	ACTIVE

5）查看帧中继永久虚电路PVC信息。

```
FR#show frame-relay pvc

PVC Statistics for interface Serial 2/0 (Frame Relay DCE)
DLCI = 26, DLCI USAGE = SWTICHED, PVC STATUS = ACTIVE , INTERFACE = Serial 2/0
   input pkts 38       output pkts 37      in bytes 988
   out bytes 1110      dropped pkts 0      in FECN pkts 0
   switched pkts 38
   in BECN pkts 0      out FECN pkts 0     out BECN pkts 0
   in DE pkts 0        out DE pkts 0

DLCI = 37, DLCI USAGE = SWTICHED, PVC STATUS = ACTIVE , INTERFACE = Serial 2/0
   input pkts 32       output pkts 32      in bytes 832
   out bytes 960       dropped pkts 0      in FECN pkts 0
   switched pkts 32
   in BECN pkts 0      out FECN pkts 0     out BECN pkts 0
   in DE pkts 0        out DE pkts 0
```

PVC Statistics for interface Serial 3/0 (Frame Relay DCE)

DLCI = 16, DLCI USAGE = SWTICHED, PVC STATUS = ACTIVE , INTERFACE = Serial 3/0
   input pkts 37   output pkts 38   in bytes 962
   out bytes 1140   dropped pkts 0   in FECN pkts 0
   switched pkts 37
   in BECN pkts 0   out FECN pkts 0   out BECN pkts 0
   in DE pkts 0   out DE pkts 0

DLCI = 36, DLCI USAGE = SWTICHED, PVC STATUS = ACTIVE , INTERFACE = Serial 3/0
   input pkts 31   output pkts 31   in bytes 806
   out bytes 930   dropped pkts 0   in FECN pkts 0
   switched pkts 31
   in BECN pkts 0   out FECN pkts 0   out BECN pkts 0
   in DE pkts 0   out DE pkts 0

PVC Statistics for interface Serial 4/0 (Frame Relay DCE)

DLCI = 17, DLCI USAGE = SWTICHED, PVC STATUS = ACTIVE , INTERFACE = Serial 4/0
   input pkts 32   output pkts 32   in bytes 832
   out bytes 960   dropped pkts 0   in FECN pkts 0
   switched pkts 32
   in BECN pkts 0   out FECN pkts 0   out BECN pkts 0
   in DE pkts 0   out DE pkts 0

DLCI = 27, DLCI USAGE = SWTICHED, PVC STATUS = ACTIVE , INTERFACE = Serial 4/0
   input pkts 31   output pkts 31   in bytes 806
   out bytes 930   dropped pkts 0   in FECN pkts 0
   switched pkts 31
   in BECN pkts 0   out FECN pkts 0   out BECN pkts 0
   in DE pkts 0   out DE pkts 0

小结

通过帧中继技术的学习，主要掌握帧中继基本配置和帧中继交换机的配置与管理工作。

第5章　网络服务应用

随着信息化的发展，局域网在人们的工作和生活中变得越来越重要。尤其是在学校的办公网络中，学校的日常教学与管理都需要网络的支撑才能很好地运转，只有充分提高网络服务性能，提高网络系统运行的效率，才能为单位网络用户提供最优质的网络服务。

本章介绍的网络服务应用主要涉及交换机与路由器的远程登录、DHCP服务、VRRP技术、VPN技术、QoS技术、组播技术和IPv6技术。

1. 交换机与路由器的远程登录

交换机与路由器的远程登录就是在交换机与路由器上启用Telnet，实现通过Telnet远程访问网络设备。

2. DHCP服务

DHCP基于Client/Server架构，是一种在网络中常用的动态编址技术。通常，DHCP服务器至少给客户端提供以下基本信息：IP地址、子网掩码、默认网关。它还可以提供其他信息，如域名服务（DNS）服务器地址和Windows Internet命名服务（WINS）服务器地址。

使用DHCP服务可以减少对上网IP地址的需求量、降低客户机的配置复杂度、减少手工配置IP地址导致的错误以及减少网络管理的工作量。

3. VRRP技术

VRRP（Virtual Router Redundancy Protocal，虚拟路由器冗余协议）是一种备份冗余解决方案，它共享多路访问介质上终端IP设备的默认网关，并进行冗余备份，从而在其中一台路由设备宕机时，备份路由设备能够及时接管转发工作，为用户提供无间断的切换，提高网络服务质量。

4. VPN技术

利用公共网络来构建私人专用网络称为虚拟私有网络（VPN，Virtual Private Network）。虚拟私有网络是指依靠ISP（Internet 服务提供商）和NSP（其他网络服务提供商），在公用网络中建立专用的数据通信网络的技术。虚拟专用网不是真的专用网络，但却能够实现专用网络的功能。

5. QoS技术

QoS（Quality of Service）又称为服务质量，是指一个网络能够利用各种各样的基础技术，向指定的网络通信提供更好的服务能力。简单地说，就是针对各种不同需求，提供不同的网络服务质量。

6. 组播技术

组播传输是指在发送者和每一位接收者之间实现点对多点的网络连接，如果一个发送者同时给多个接收者传输相同的数据，也只需复制一份相同的数据包。

组播协议允许将一台主机发送的数据通过网络路由器和交换机复制到多个加入此组播的主机，是一对多的通信方式。在组播网络中，即使组播用户数量成倍增长，骨干网络中网络带宽也无需增加。

7. IPv6技术

IPv6（Internet Protocol Version 6）是IETF（Internet Engineering Task Force，互联网工程任务组）设计的用于替代现行版本IP协议（IPv4）的新一代IP协议。IPv6具有长达128位的地址空间，可以彻底解决IPv4地址不足的问题，除此之外，IPv6还采用了分级地址模式、高效IP包头、服务质量、主机地址自动配置、认证和加密等许多技术。

5.1　Telnet服务

问题描述

该校网络中有许多网络设备，为了便于管理，提高工作效率，需要在网络设备上开启Telnet服务。由于远程登录的不安全性，为加强其安全性能，因此需要对line线路进行保护。

问题分析

配置交换机与路由器的Telnet服务，主要涉及二层交换机的远程登录、三层交换机的远程登录与路由器的远程登录。

1. 配置二层交换机的远程登录

步骤1　配置telnet密码。参数0表示可以输入一个明文口令，5则表示需要输入一个已经加密的口令。level 1表示普通用户级别，即是远程登录级别。encrypted-password表示telnet密码。

switch(config)#**enable secret level 1 0 | 5** *encrypted-password*

步骤2　配置enable密码。level 15表示最高用户级别，即是enable登录级别。encrypted-password表示enable密码。

switch(config)#**enable secret level 15 0|5** *encrypted-password*

2. 配置三层交换机的远程登录

步骤1　配置enable密码。encrypted-password口令字符串表示enable密码。

switch(config)#**enable password 0** *encrypted-password*

步骤2　进入线程配置模式。

switch(config-line)#**line vty 0 4**

步骤3　配置telnet密码。encrypted-password口令字符串表示telnet密码。

switch(config-line)#**password 0** *encrypted-password*

第5章 网络服务应用

步骤4 启用line线路保护,开启telnet的用户名密码验证。
switch(config-line)#**login**

3. 配置路由器的远程登录

步骤1 配置enable密码。encrypted-password口令字符串表示enable密码。
router(config)#**enable password 0** *encrypted-password*
步骤2 进入线程配置模式。
router(config-line)#**line vty 0 4**
步骤3 配置telnet密码。encrypted-password口令字符串表示telnet密码。
router(config-line)#**password 0** *encrypted-password*
步骤4 启用line线路保护,开启telnet的用户名密码验证。
switch(config-line)#**login**

4. 配置限制对路由器的VTY访问

步骤1 配置访问控制列表。
router(config)#**access-list** *list-number* {**permit** | **deny**} {*source source-wildcard*}
步骤2 进入线程配置模式,限制对路由器的VTY访问。
router(config-line)#**line vty 0 4**
router(config-line)#**access-class** *list-number* **in**

任 务 单

1	交换机的远程登录
2	路由器的远程登录

根据任务单的安排完成任务。

任务1:交换机的远程登录

任务实施

1. 任务描述及网络拓扑设计

要求在二层交换机和三层交换机上配置远程登录,网络管理员可以远程以Telnet方式登录配置。绘制拓扑结构图,如图5-1所示。

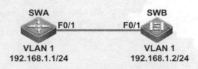

图5-1 交换机远程登录配置

2. 网络设备配置

(1)配置管理IP地址

1)在SWA交换机上配置管理IP地址。

```
SWA(config)#interface vlan 1
SWA(config-VLAN 1)#ip address 192.168.1.1 255.255.255.0
```
2）在SWB交换机上配置管理IP地址。
```
SWB(config)#interface vlan 1
SWB(config-VLAN 1)#ip address 192.168.1.2 255.255.255.0
```
（2）在SWA三层交换机上配置远程登录

SWA(config)#enable password 0 123456	#配置enable密码
SWA(config)#line vty 0 4	#进入线程配置模式
SWA(config-line)#password 0 654321	#配置Telnet密码
SWA(config-line)#login	#启用Telnet的用户名密码验证

（3）在SWB三层交换机上配置远程登录。

SWB(config)#enable secret level 1 0 654321	#配置Telnet密码
SWB(config)#enable secret level 15 0 123456	#配置enable密码
SWB(config)#line vty 0 4	#进入线程配置模式
SWB(config-line)#login	#启用Telnet的用户名密码验证

（4）使用Telnet远程登录验证

1）在交换机SWA上远程登录到交换机SWB。
```
SWA#telnet 192.168.1.2
Trying 192.168.1.2, 23...
User Access Verification
Password:
#提示输入Telnet密码，输入设置的密码为654321
SWB>enable
Password:
#提示输入enable密码，输入设置的密码为123456
SWB#
#现在已经进入SWB交换机特权模式，可以正常地进行配置
```
2）在交换机SWB上远程登录到交换机SWA。
```
SWB#telnet 192.168.1.1
Trying 192.168.1.1, 23...
User Access Verification
Password:
#提示输入Telnet密码，输入设置的密码为654321
SWA>enable
Password:
#提示输入enable密码，输入设置的密码为123456
SWA#
```

任务2：路由器的远程登录

任务实施

1. 任务描述及网络拓扑设计

要求在RA与RB路由器上配置远程登录，网络管理员可以远程以Telnet方式登录配置。其中RB路由器只允许192.168.1.6/24用户可以Telnet远程访问。绘制拓扑结构图，如图5-2所示。

第5章 网络服务应用

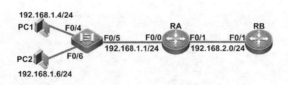

图5-2 路由器远程登录配置

2. 网络设备配置

（1）路由器基本配置

1）在RA交换机上配置管理IP地址。

RA(config)#interface fastEthernet 0/0
RA(config-if-FastEthernet 0/0)#ip address 192.168.1.1 255.255.255.0
RA(config-if-FastEthernet 0/0)#exit
RA(config)#interface fastEthernet 0/1
RA(config-if-FastEthernet 0/1)#ip address 192.168.2.1 255.255.255.0

2）在RB交换机上配置管理IP地址。

RB(config)#interface fastEthernet 0/1
RB(config-if-FastEthernet 0/1)#ip address 192.168.2.2 255.255.255.0

（2）配置路由，实现RA与RB之间的互通

1）在RA交换机上配置默认路由。

RA(config)#ip route 0.0.0.0 0.0.0.0 192.168.2.2   #配置到RB的默认路由

2）在RB交换机上配置默认路由。

RB(config)#ip route 0.0.0.0 0.0.0.0 192.168.2.1   #配置到RA的默认路由

（3）在RA与RB上配置Telnet

1）在RA交换机上配置Telnet。

RA(config)#enable password 0 123456   #配置enable密码
RA(config)#line vty 0 4   #进入线程配置模式
RA(config-line)#password 654321   #配置Telnet密码
RA(config-line)#login   #设置Telnet登录时身份验证

2）在RB交换机上配置Telnet。

RB(config)#enable password 0 123456   #配置enable密码
RB(config)#line vty 0 4   #进入线程配置模式
RB(config-line)#password 654321   #配置Telnet密码
RB(config-line)#login   #设置Telnet登录时身份验证

（4）配置访问控制列表

RB(config)#access-list 10 permit host 192.168.1.6   #配置访问列表
RB(config)#lin vty 0 4   #进入线程配置模式
RB(config-line)#access-class 10 in   #进行VTY访问限制
RB(config-line)#exit

（5）验证测试

将PC1的IP地址设置为192.168.1.4/24，网关设置为192.168.1.1；将PC2的IP地址设置为192.168.1.6/24，网关设置为192.168.1.1；分别远程登录路由器RA与RB，观察效果。

1）在PC1和PC2上远程登录路由器RA，如图5-3所示。

图5-3　PC1和PC2远程登录RA

2）在RA路由器上远程登录路由器RB。
RA#telnet 192.168.2.2
Trying 192.168.2.2, 23…
% Destination unreachable; gateway or host down

3）在PC2上远程登录路由器RB，如图5-4所示。

图5-4　PC2远程登录RB

小结

通过交换机与路由器远程登录的学习，主要掌握交换机的远程登录、路由器的远程登录、VTY访问限制等配置与管理工作。

5.2　DHCP服务

问题描述

为了减少内部用户对上网IP地址的需求量、降低客户机的配置复杂度、减少手工配置IP地址导致的错误以及减少网络管理的工作量，需要为内部用户主机动态分配IP地址。可以在Windows服务器或者在汇聚层交换机上为主校区VLAN 10和VLAN 20用户部署DHCP

第5章 网络服务应用

服务，分配IP地址时，需要为用户主机配置网关和DNS服务器。使用Windows服务器部署DHCP服务时，需要配置DHCP中继。

问题分析

配置DHCP服务时，主要涉及DHCP地址池配置和DHCP中继配置。

1. 配置DHCP地址池

步骤1 启用DHCP服务器。
router(config)#**service dhcp**
步骤2 创建DHCP地址池。参数pool-name表示地址池的名称。
router(config)#**ip dhcp pool** *pool-name*
步骤3 配置地址池范围和掩码。network-number表示网络地址，mask表示子网掩码。
router(dhcp-config)#**network** *network-number mask*
步骤4 配置地址租约。days表示租期的时间，以天为单位。Hours表示租期的时间，以小时为单位，定义小时数前必须定义天数。Minutes表示租期的时间，以分钟为单位，定义分钟数前必须定义天数和小时数。infinite表示没有限制的租期。
router(dhcp-config)#**lease** {**days**[*hours*[*minutes*]] | **infinite** }
步骤5 配置默认网关。address表示默认网关。
router(dhcp-config)#**default-router** *address*
步骤6 配置域名。参数domain-name表示分配给客户端的域名后缀。
router(dhcp-config)#**domain-name** *domain-name*
步骤7 配置DNS服务器。可以配置多个服务器地址，最多配置8个。
router(dhcp-config)#**dns-server** *address1* [*address2*…*address8*]
步骤8 配置WINS服务器。WINS是网络名称解析服务，它将计算机的NetBIOS名称转换为相应的IP地址。
router(dhcp-config)#**netbios-name-server** *address1* [*address2*…*address8*]
步骤9 配置NetBIOS节点类型。参数type表示客户端NetBIOS节点类型，可以为b-node、p-node、m-node和h-mode。Windows客户端在使用WINS进行NetBIOS名称解析时，可以使用4种不同的方式，每一种NetBIOS节点类型表示一种方式。b-node表示广播节点类型，该节点类型的客户端使用广播的方式对NetBIOS名称进行解析。p-node表示对等节点类型，该节点类型的客户端使用WINS服务器（单播）对NetBIOS名称进行解析。m-node表示混合节点类型，该节点类型的客户端先使用广播方式进行解析，如果解析失败，则使用WINS服务器（单播）对NetBIOS名称进行解析。h-node表示复合节点类型，该节点类型的客户端先使用WINS服务器（单播）进行解析，如果解析失败，则使用广播方式对NetBIOS名称进行解析。
router(dhcp-config)#**netbios-node-type** *type*

2. 配置静态地址绑定

步骤1 配置静态地址。mask表示子网掩码，如果不指定将使用默认的子网掩码。
router(dhcp-config)#**host** *address* [*mask*]
步骤2 配置客户端硬件地址。
router(dhcp-config)#**hardware-address** *hardware-address*

3. 配置排除地址

配置排除地址时，参数start-address和end-address表示排除地址范围的起始地址和结束地址。

router(config)#**ip dhcp excluded-address** [*start-address end-address*]

4. 配置DHCP中继代理

步骤1 启用DHCP服务器。
router(config)#**service dhcp**
步骤2 配置DHCP中继代理。参数address表示中继地址，即DHCP服务器的地址。
router(config)#**ip helper-address** *address*

任 务 单

1	配置DHCP服务
2	配置DHCP中继代理

解决步骤

根据任务单的安排完成任务。

任务1：配置DHCP服务

任务实施

1. 任务描述及网络拓扑设计

要求在三层交换机SW1上配置DHCP服务，分别为VLAN 10和VLAN 20用户分配IP地址，需要为主机配置默认网关和DNS服务器。服务群中DNS服务器的地址为192.168.100.88，公网提供的DNS服务器为68.1.1.6，域名为abc.com.cn，并且每个VLAN中的主机位是160～200的地址不允许分配给客户端。绘制拓扑结构图，如图5-5所示。

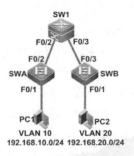

图5-5 DHCP服务配置

2. 网络设备配置

（1）SW1交换机配置
1）在SW1交换机上创建VLAN和SVI接口。

```
SW1(config)#vlan 10
SW1(config-vlan)#exit
SW1(config)#vlan 20
SW1(config-vlan)#exit
SW1(config)#interface vlan 10
SW1(config-VLAN 10)#ip address 192.168.10.1 255.255.255.0
SW1(config-VLAN 10)#exit
SW1(config)#interface vlan 20
SW1(config-VLAN 20)#ip address 192.168.20.1 255.255.255.0
SW1(config-VLAN 20)#exit
SW1(config)#interface range fastEthernet 0/2-3
SW1(config-if-range)#switchport mode trunk
```

2）配置DHCP地址池。

```
SW1(config)#service dhcp                                    #启用DHCP服务
SW1(config)#ip dhcp pool vlan10                             #设定地址池名称
SW1(dhcp-config)#network 192.168.10.0 255.255.255.0         #配置DHCP地址池
SW1(dhcp-config)#dns-server 192.168.100.88 68.1.1.6         #配置DNS地址
SW1(dhcp-config)#domain-name abc.com.cn                     #配置域名
SW1(dhcp-config)#default-router 192.168.10.1                #配置网关
SW1(dhcp-config)#exit
SW1(config)#ip dhcp excluded-address 192.168.10.160 192.168.10.200   #配置排除地址
SW1(config)#ip dhcp pool vlan20                             #设定地址池名称
SW1(dhcp-config)#network 192.168.20.0 255.255.255.0         #配置DHCP地址池
SW1(dhcp-config)#dns-server 192.168.100.88 68.1.1.6         #配置DNS地址
SW1(dhcp-config)#domain-name abc.com.cn                     #配置域名
SW1(dhcp-config)#default-router 192.168.20.1                #配置网关
SW1(dhcp-config)#exit
SW1(config)#ip dhcp excluded-address 192.168.20.160 192.168.20.200   #配置排除地址
```

（2）SWA交换机配置

```
SWA(config)#vlan 10
SWA(config-vlan)#exit
SWA(config)#interface fastEthernet 0/1
SWA(config-if)#switchport access vlan 10
SWA(config-if)#exit
SWA(config)#interface fastEthernet 0/2
SWA(config-if)#switchport mode trunk
```

（3）SWB交换机配置

```
SWB(config)#vlan 20
SWB(config-vlan)#exit
SWB(config)#interface fastEthernet 0/1
SWB(config-if)#switchport access vlan 20
SWB(config-if)#exit
SWB(config)#interface fastEthernet 0/3
SWB(config-if)#switchport mode trunk
```

（4）验证测试

1）将PC1连接到SWA交换机的F0/1端口，来获取IP地址。本地连接中地址配置选项设置为"自动获取IP地址"，在DOS命令行配置界面使用命令ipconfig/all查看主机IP，如图5-6所示，从图中可以看到主机通过自动获取IP的方式得到指定VLAN 10网段的IP。

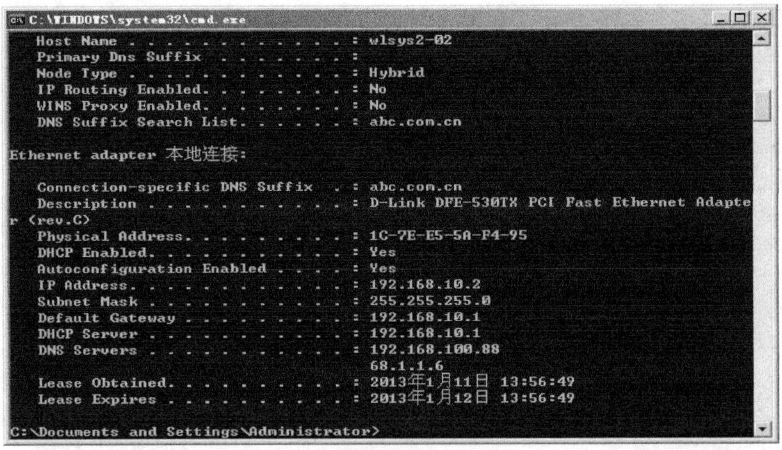

图5-6 设置自动获取IP地址

2）将PC2连接到SWB交换机的F0/1端口，来获取IP地址。本地连接中地址配置选项设置为"自动获取IP地址"，在DOS命令行配置界面使用命令ipconfig/all查看主机IP，如图5-7所示，从图中可以看到主机通过自动获取IP的方式得到指定VLAN 20网段的IP。

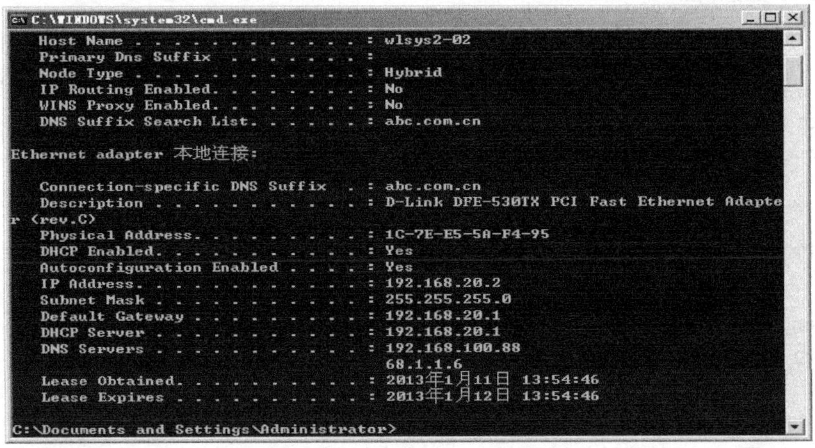

图5-7 设置自动获取IP地址

任务2：配置DHCP中继代理

任务实施

1. 任务描述及网络拓扑设计

使用Windows服务器部署DHCP服务，分别为VLAN 10和VLAN 20用户分配IP地址，需要为主机配置默认网关和DNS服务器。其中VLAN 30是服务器群，为静态指定的IP地址，服务器群中DHCP服务器的地址为192.168.30.100，DNS服务器的地址为192.168.30.200。公网提供的DNS服务器为68.1.1.6，域名为aaa.com.cn。由于DHCP客户端和DHCP服务器不在同一个子网，为了使客户端能够成功获得IP地址，需要使用DHCP中继代理。绘制拓扑结构图，如图5-8所示。

第5章 网络服务应用

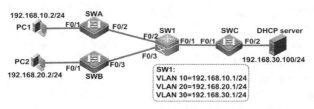

图5-8 DHCP中继配置

2. 网络设备配置

（1）在SW1交换机上创建VLAN和SVI接口

SW1(config)#vlan 10
SW1(config-vlan)#exit
SW1(config)#vlan 20
SW1(config-vlan)#exit
SW1(config)#vlan 30
SW1(config-vlan)#exit
SW1(config)#interface vlan 10
SW1(config-VLAN 10)#ip address 192.168.10.1 255.255.255.0
SW1(config-VLAN 10)#exit
SW1(config)#interface vlan 20
SW1(config-VLAN 20)#ip address 192.168.20.1 255.255.255.0
SW1(config-VLAN 20)#exit
SW1(config)#interface vlan 30
SW1(config-VLAN 30)#ip address 192.168.30.1 255.255.255.0
SW1(config-VLAN 30)#exit
SW1(config)#interface range fastEthernet 0/1-3
SW1(config-if-range)#switchport mode trunk

（2）在SWA交换机上创建VLAN和端口隔离

SWA(config)#vlan 10
SWA(config-vlan)#exit
SWA(config)#interface fastEthernet 0/1
SWA(config-FastEthernet 0/1)#switchport access vlan 10
SWA(config-FastEthernet 0/1)#exit
SWA(config)#interface fastEthernet 0/2
SWA(config-FastEthernet 0/2)#switchport mode trunk

（3）在SWB交换机上创建VLAN和端口隔离

SWB(config)#vlan 20
SWB(config-vlan)#exit
SWB(config)#interface fastEthernet 0/1
SWB(config-FastEthernet 0/1)#switchport access vlan 20
SWB(config-FastEthernet 0/1)#exit
SWB(config)#interface fastEthernet 0/3
SWB(config-FastEthernet 0/3)#switchport mode trunk
SWB(config-FastEthernet 0/3)#exit

（4）在SWC交换机上创建VLAN和端口隔离

SWC(config)#vlan 10
SWC(config-vlan)#exit
SWC(config)#vlan 20

```
SWC(config-vlan)#exit
SWC(config)#vlan 30
SWC(config-vlan)#exit
SWC(config)#interface fastEthernet 0/1
SWC(config-if)#switchport mode trunk
SWC(config-if)#exit
SWC(config)#interface fastEthernet 0/2
SWC(config-if)#switchport access vlan 30
SWC(config-if)#exit
```

（5）配置DHCP中继代理

```
SW1(config)#service dhcp                              #启用DHCP服务
SW1(config)#ip helper-address 192.168.30.100          #配置DHCP中继
```

（6）验证测试

1）将PC1连接到SWA交换机的F0/1端口，来获取IP地址。本地连接中地址配置选项设置为"自动获取IP地址"，在DOS命令行配置界面使用命令ipconfig/all查看主机IP，如图5-9所示，从图中可以看到主机通过自动获取IP的方式得到指定VLAN 10网段的IP。

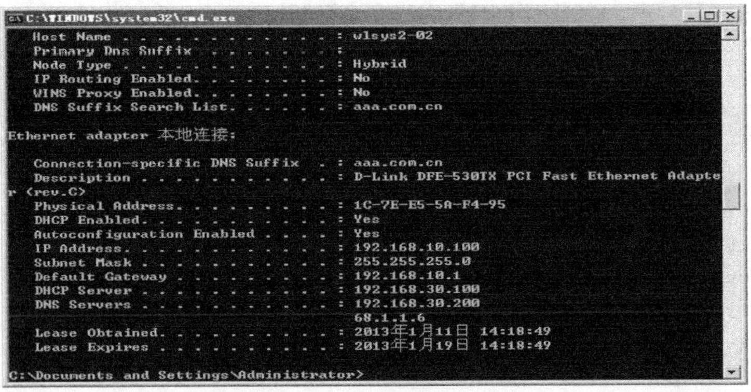

图5-9　PC1设置自动获取IP地址

2）将PC2连接到SWB交换机的F0/1端口，来获取IP地址。本地连接中地址配置选项设置为"自动获取IP地址"，在DOS命令行配置界面使用命令ipconfig/all查看主机IP，如图5-10所示，从图中可以看到主机通过自动获取IP的方式得到指定VLAN 20网段的IP。

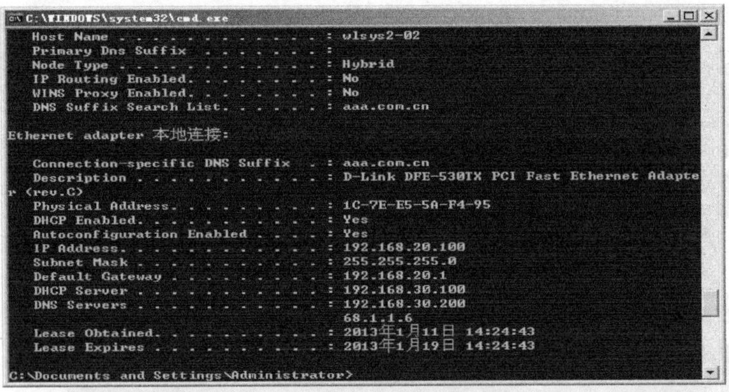

图5-10　PC2设置自动获取IP地址

小结

通过DHCP服务的学习,主要掌握DHCP地址和DHCP中继代理等配置与管理工作。

5.3 VRRP技术

 问题描述

为了实现该校分校区三层链路的冗余和负载均衡,需要在网络中使用VRRP协议。要求设置一台三层交换机SW3为VLAN 40和VLAN 50的活跃路由器,另一台三层交换机SW4为备份路由器;设置SW4为VLAN 60的活跃路由器,SW3为备份路由器,并要求使用SVI接口地址作为虚拟路由器的IP地址。

问题分析

配置VRRP主要涉及VRRP单备份组配置、VRRP多备份组配置和基于SVI的VRRP备份组配置。

1. VRRP基本配置

步骤1 在接口模式下,配置VRRP组。参数group-number是VRRP组的编号,其取值范围是1～255;ip-address是VRRP组的虚拟IP地址,虚拟IP地址可以是该子网中使用的IP地址,也可以是某台VRRP路由器接口的IP地址,即是IP地址拥有者;secondary表示该VRRP组配置辅助IP地址。

router(config-if)#**vrrp** *group-number* **ip** *ip-address* [**secondary**]

步骤2 配置VRRP优先级。参数group-number是VRRP组的编号;number表示优先级数值,取值范围是1～254,默认是100。

router(config-if)#**vrrp** *group-number* **priority** *number*

2. 调整和优化VRRP配置

步骤1 配置VRRP接口跟踪。参数interface表示被跟踪的接口;priority-decrement表示VRRP发现被跟踪接口不可用后,所降低的优先级数值,默认为10,当跟踪接口恢复后,优先级也将恢复到原先的值。

router(config-if)#**vrrp** *group-number* **track** *interface* [*priority-decrement*]

步骤2 配置VRRP抢占模式。参数group-number是VRRP组的编号;delay-time表示抢占的延迟时间,即发送通告报文前等待的时间,时间为秒(s),取值范围为1～255,如果不配置延迟时间,默认值为0s。

router(config-if)#**vrrp** *group-number* **preempt** [**delay** *delay-time*]

步骤3 配置VRRP定时器。参数group-number是VRRP组的编号;advertise-interval

表示通告报文的发送间隔，单位为秒，取值范围为1～255，默认值为1s。

router(config-if)#**vrrp** *group-number* **timers advertise** *advertise-interval*

步骤4 配置VRRP验证。参数string表示明文密码。

router(config-if)#**vrrp** *group-number* **authentication** *string*

任 务 单

1	VRRP单备份组
2	VRRP多备份组
3	基于SVI的VRRP备份组

解决步骤

根据任务单的安排完成任务。

任务1：配置VRRP单备份组

任务实施

1. 任务描述及网络拓扑设计

将RB和RC路由器配置到一个VRRP组中，每一个VRRP组虚拟出一台虚拟路由器，作为网络中主机的网关。绘制拓扑结构图，如图5-11所示。

图5-11 VRRP单备份组

2. 网络设备配置

（1）在路由器上配置IP地址

1）在RA路由器上配置IP地址。

RA(config)#interface serial 2/0
RA(config-if-Serial 2/0)#ip address 50.1.1.2 255.255.255.0
RA(config-if-Serial 2/0)#exit
RA(config)#interface serial 3/0
RA(config-if-Serial 3/0)#ip address 51.1.1.2 255.255.255.0
RA(config-if-Serial 3/0)#exit

2）在RB路由器上配置IP地址。

RB(config)#interface fastEthernet 0/0
RB(config-if-FastEthernet 0/0)#ip address 52.1.1.1 255.255.255.0
RB(config-if-FastEthernet 0/0)#exit
RB(config)#interface serial 2/0
RB(config-if-Serial 2/0)#ip address 50.1.1.1 255.255.255.0
RB(config-if-Serial 2/0)#exit

3）在RC路由器上配置IP地址。

RC(config)#interface serial 3/0
RC(config-if-Serial 3/0)#ip address 51.1.1.1 255.255.255.0
RC(config-if-Serial 3/0)#exit
RC(config)#interface fastEthernet 0/0
RC(config-if-FastEthernet 0/0)#ip address 52.1.1.2 255.255.255.0
RC(config-if-FastEthernet 0/0)#exit

（2）在路由器上配置路由

1）在RA路由器上配置RIP路由。

RA(config)#router rip
RA(config-router)#no auto-summary
RA(config-router)#version 2
RA(config-router)#network 51.1.1.0
RA(config-router)#network 50.1.1.0

2）在RB路由器上配置RIP路由。

RB(config)#router rip
RB(config-router)#no auto-summary
RB(config-router)#version 2
RB(config-router)#network 52.1.1.0
RB(config-router)#network 50.1.1.0

3）在RC路由器上配置RIP路由。

RC(config)#router rip
RC(config-router)#no auto-summary
RC(config-router)#version 2
RC(config-router)#network 51.1.1.0
RC(config-router)#network 52.1.1.0

（3）在RB和RC路由器上配置VRRP

1）在RB路由器上配置VRRP。

RB(config)#interface fastEthernet 0/0
RB(config-if-FastEthernet 0/0)#vrrp 1 ip 52.1.1.254   #启用VRRP进程
RB(config-if-FastEthernet 0/0)#vrrp 1 priority 120   #定义接口VRRP优先级
RB(config-if-FastEthernet 0/0)#vrrp 1 preempt   #启用抢占模式
RB(config-if-FastEthernet 0/0)#vrrp 1 track serial 2/0 100   #配置接口跟踪
RB(config-if-FastEthernet 0/0)#exit

2）在RC路由器上配置VRRP。

RC(config)#interface fastEthernet 0/0
RC(config-if-FastEthernet 0/0)#vrrp 1 ip 52.1.1.254   #启用VRRP进程
RC(config-if-FastEthernet 0/0)#vrrp 1 priority 100   #定义接口VRRP优先级
RC(config-if-FastEthernet 0/0)#vrrp 1 preempt   #启用抢占模式
RC(config-if-FastEthernet 0/0)#exit

（4）验证测试

1）在RB上查看VRRP。

RB#show vrrp brief
Interface   Grp   Pri   timer   Own   Pre   State   Master addr   Group addr
FastEthernet 0/0 1   120   3   -   P   Master   52.1.1.1   52.1.1.254

从show命令的输出结果可以看到，RB路由器在VRRP组1中，优先级为120，状态为Master路由器。

2)在RC上查看VRRP。

```
RC#show vrrp brief
Interface        Grp  Pri  timer  Own  Pre  State   Master addr  Group addr
FastEthernet 0/0  1   100   3      -    P   Backup  52.1.1.1     52.1.1.254
```

从show命令的输出结果可以看到，RC路由器在VRRP组1中，优先级为100，状态为Backup路由器。

任务2：配置VRRP多备份组

任务实施

1. 任务描述及网络拓扑设计

将RB和RC路由器配置到VRRP组1和VRRP组2中，并且在不同的VRRP组中担任不同的角色，从而保证接入线路上的路由器都能够承担转发任务。绘制拓扑结构图，如图5-12所示。

图5-12　VRRP多备份组

2. 网络设备配置

（1）在路由器上配置IP地址

1）在RA路由器上配置IP地址。

RA(config)#interface serial 2/0
RA(config-if-Serial 2/0)#ip address 50.1.1.2 255.255.255.0
RA(config-if-Serial 2/0)#exit
RA(config)#interface serial 3/0
RA(config-if-Serial 3/0)#ip address 51.1.1.2 255.255.255.0
RA(config-if-Serial 3/0)#exit

2）在RB路由器上配置IP地址。

RB(config)#interface fastEthernet 0/0
RB(config-if-FastEthernet 0/0)#ip address 52.1.1.1 255.255.255.0
RB(config-if-FastEthernet 0/0)#exit
RB(config)#interface serial 2/0
RB(config-if-Serial 2/0)#ip address 50.1.1.1 255.255.255.0
RB(config-if-Serial 2/0)#exit

3）在RC路由器上配置IP地址。

RC(config)#interface serial 3/0
RC(config-if-Serial 3/0)#ip address 51.1.1.1 255.255.255.0
RC(config-if-Serial 3/0)#exit
RC(config)#interface fastEthernet 0/0
RC(config-if-FastEthernet 0/0)#ip address 52.1.1.2 255.255.255.0
RC(config-if-FastEthernet 0/0)#exit

(2) 在路由器上配置路由

1) 在RA路由器上配置RIP路由。

RA(config)#router rip
RA(config-router)#no auto-summary
RA(config-router)#version 2
RA(config-router)#network 51.1.1.0
RA(config-router)#network 50.1.1.0
RA(config-router)#exit

2) 在RB路由器上配置RIP路由。

RB(config)#router rip
RB(config-router)#no auto-summary
RB(config-router)#version 2
RB(config-router)#network 52.1.1.0
RB(config-router)#network 50.1.1.0

3) 在RC路由器上配置RIP路由。

RC(config)#router rip
RC(config-router)#no auto-summary
RC(config-router)#version 2
RC(config-router)#network 51.1.1.0
RC(config-router)#network 52.1.1.0

(3) 在RB和RC路由器上配置VRRP

1) 在RB路由器上配置VRRP。

RB(config)#interface fastEthernet 0/0
RB(config-if-FastEthernet 0/0)#vrrp 1 ip 52.1.1.254   #启用VRRP进程
RB(config-if-FastEthernet 0/0)#vrrp 1 priority 120   #定义接口VRRP优先级
RB(config-if-FastEthernet 0/0)#vrrp 1 preempt   #启用抢占模式
RB(config-if-FastEthernet 0/0)#vrrp 1 track serial 2/0 100   #配置接口跟踪和降低优先级
RB(config-if-FastEthernet 0/0)#vrrp 2 ip 52.1.1.253   #启用VRRP进程
RB(config-if-FastEthernet 0/0)#vrrp 2 preempt
RB(config-if-FastEthernet 0/0)#vrrp 2 priority 100   #定义接口VRRP优先级

2) 在RC路由器上配置VRRP。

RC(config)#interface fastEthernet 0/0
RC(config-if-FastEthernet 0/0)#vrrp 1 ip 52.1.1.254   #启用VRRP进程
RC(config-if-FastEthernet 0/0)#vrrp 1 priority 100   #定义接口VRRP优先级
RC(config-if-FastEthernet 0/0)#vrrp 1 preempt   #启用抢占模式
RC(config-if-FastEthernet 0/0)#vrrp 2 ip 52.1.1.253   #启用VRRP进程
RC(config-if-FastEthernet 0/0)#vrrp 2 preempt   #启用抢占模式
RC(config-if-FastEthernet 0/0)#vrrp 2 priority 120   #定义接口VRRP优先级
RC(config-if-FastEthernet 0/0)#vrrp 2 track serial 3/0 100   #配置接口跟踪和降低优先级

(4) 验证测试

1) 在RB上查看VRRP。

RB#show vrrp brief

Interface	Grp	Pri	timer	Own	Pre	State	Master addr	Group addr
FastEthernet 0/0	1	120	3	-	P	Master	52.1.1.1	52.1.1.254
FastEthernet 0/0	2	100	3	-	P	Backup	52.1.1.2	52.1.1.253

从show命令的输出结果可以看到，RB路由器在VRRP组1中，优先级为120，状态为Master路由器；在VRRP组2中，优先级为100，状态为Backup路由器。

2)在RC上查看VRRP。

```
RC#show vrrp brief
Interface        Grp  Pri  timer  Own  Pre  State   Master addr  Group addr
FastEthernet 0/0  1   100   3     -    P    Backup  52.1.1.1     52.1.1.254
FastEthernet 0/0  2   120   3     -    P    Master  52.1.1.2     52.1.1.253
```

从show命令的输出结果可以看到，RC路由器在VRRP组1中，优先级为100，状态为Backup路由器；在VRRP组2中，优先级为120，状态为Master路由器。

任务3：配置基于SVI的VRRP备份组

任务实施

1. 任务描述及网络拓扑设计

配置基于SVI的VRRP多备份组实现VRRP的负载均衡。汇聚层交换机SWA上创建有VLAN 10、VLAN20、VLAN30和VLAN40，各VLAN的IP地址分别为192.168.10.1/24、192.168.20.1/24、192.168.30.1/24和192.168.40.1/24。汇聚层交换机SWB上创建有VLAN 10、VLAN20、VLAN30和VLAN40，各VLAN的IP地址分别为192.168.10.2/24、192.168.20.2/24、192.168.30.2/24和192.168.40.2/24。将SWA和SWB三层交换机分别分配到VRRP组10和20中，同一台三层交换机在不同的VRRP组中承担不同的角色，使得所有的三层交换机都承担数据转发任务。绘制拓扑结构图，如图5-13所示。

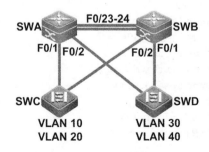

图5-13　VRRP多备份组

2. 网络设备配置

（1）在交换机上创建VLAN

1）在SWA交换机上创建VLAN。

```
SWA(config)#vlan 10
SWA(config-vlan)#exit
SWA(config)#vlan 20
SWA(config-vlan)#exit
SWA(config)#vlan 30
SWA(config-vlan)#exit
SWA(config)#vlan 40
```

2）在SWB交换机上创建VLAN。

```
SWB(config)#vlan 10
SWB(config-vlan)#exit
```

```
SWB(config)#vlan 20
SWB(config-vlan)#exit
SWB(config)#vlan 30
SWB(config-vlan)#exit
SWB(config)#vlan 40
```

3）在SWC交换机上创建VLAN。

```
SWC(config)#vlan 10
SWC(config-vlan)#exit
SWC(config)#vlan 20
SWC(config-vlan)#exit
SWC(config)#vlan 30
SWC(config-vlan)#exit
SWC(config)#vlan 40
```

4）在SWD交换机上创建VLAN。

```
SWD(config)#vlan 10
SWD(config-vlan)#exit
SWD(config)#vlan 20
SWD(config-vlan)#exit
SWD(config)#vlan 30
SWD(config-vlan)#exit
SWD(config)#vlan 40
```

（2）在交换机上配置IP地址

1）在SWA交换机上配置IP地址。

```
SWA(config)#interface vlan 10
SWA(config-VLAN 10)#ip address 192.168.10.1 255.255.255.0
SWA(config-VLAN 10)#exit
SWA(config)#interface vlan 20
SWA(config-VLAN 20)#ip address 192.168.20.1 255.255.255.0
SWA(config-VLAN 20)#exit
SWA(config)#interface vlan 30
SWA(config-VLAN 30)#ip address 192.168.30.1 255.255.255.0
SWA(config-VLAN 30)#exit
SWA(config)#interface vlan 40
SWA(config-VLAN 40)#ip address 192.168.40.1 255.255.255.0
SWA(config-VLAN 40)#exit
```

2）在SWB交换机上配置IP地址。

```
SWB(config)#interface vlan 10
SWB(config-VLAN 10)#ip address 192.168.10.2 255.255.255.0
SWB(config-VLAN 10)#exit
SWB(config)#interface vlan 20
SWB(config-VLAN 20)#ip address 192.168.20.2 255.255.255.0
SWB(config-VLAN 20)#exit
SWB(config)#interface vlan 30
SWB(config-VLAN 30)#ip address 192.168.30.2 255.255.255.0
SWB(config-VLAN 30)#exit
SWB(config)#interface vlan 40
SWB(config-VLAN 40)#ip address 192.168.40.2 255.255.255.0
SWB(config-VLAN 40)#exit
```

（3）配置Trunk及链路聚合

1）在SWA交换机上配置Trunk及链路聚合。

SWA(config)#interface range fastEthernet 0/1-2
SWA(config-if-range)#switchport mode trunk
SWA(config-if-range)#exit
SWA(config)#interface range fastEthernet 0/23-24
SWA(config-if-range)#port-group 1
SWA(config-if-range)#exit
SWA(config)#interface agregateport 1
SWA(config-if)#switchport mode trunk
SWA(config-if)#exit

2）在SWB交换机上配置Trunk及链路聚合。

SWB(config)#interface range fastEthernet 0/1-2
SWB(config-if-range)#switchport mode trunk
SWB(config-if-range)#exit
SWB(config)#interface range fastEthernet 0/23-24
SWB(config-if-range)#port-group 1
SWB(config-if-range)#exit
SWB(config)#interface agregateport 1
SWB(config-if)#switchport mode trunk
SWB(config-if)#exit

3）在SWC交换机上配置Trunk。

SWC(config)#interface range fastEthernet 0/1-2
SWC(config-if-range)#switchport mode trunk
SWC(config-if-range)#exit

4）在SWD交换机上配置Trunk。

SWD(config)#interface range fastEthernet 0/1-2
SWD(config-if-range)#switchport mode trunk
SWD(config-if-range)#exit

（4）在SWA和SWB交换机上配置SVI的VRRP

1）在SWA交换机上配置VRRP。

SWA(config)#interface vlan 10
SWA(config-VLAN 10)#vrrp 10 ip 192.168.10.254
SWA(config-VLAN 10)#vrrp 10 preempt
SWA(config-VLAN 10)#vrrp 10 priority 120
SWA(config-VLAN 10)#exit
SWA(config)#interface vlan 20
SWA(config-VLAN 20)#vrrp 20 ip 192.168.20.254
SWA(config-VLAN 20)#vrrp 20 preempt
SWA(config-VLAN 20)#vrrp 20 priority 100
SWA(config-VLAN 20)#exit

2）在SWB交换机上配置VRRP。

SWB(config)#interface vlan 10
SWB(config-VLAN 10)#vrrp 10 ip 192.168.10.254
SWB(config-VLAN 10)#vrrp 10 preempt
SWB(config-VLAN 10)#vrrp 10 priority 100
SWB(config-VLAN 10)#exit
SWB(config)#interface vlan 20

```
SWB(config-VLAN 20)#vrrp 20 ip 192.168.20.254
SWB(config-VLAN 20)#vrrp 20 preempt
SWB(config-VLAN 20)#vrrp 20 priority 120
SWB(config-VLAN 20)#exit
```

（5）验证测试

1）在SWA上查看VRRP。

```
SWA#show vrrp brief
Interface        Grp  Pri   timer  Own  Pre  State    Master addr     Group addr
VLAN 10    10   120   3      -    P    Master   192.168.10.1    192.168.10.254
VLAN 20    20   100   3      -    P    Backup   192.168.20.2    192.168.20.254
```

从show命令的输出结果可以看到，SWA路由器在VRRP组10中，优先级为120，状态为Master路由器；在VRRP组20中，优先级为100，状态为Backup路由器。

2）在SWB上查看VRRP。

```
SWB#show vrrp brief
Interface        Grp  Pri   timer  Own  Pre  State    Master addr     Group addr
VLAN 10    10   100   3      -    P    Backup   192.168.10.1    192.168.10.254
VLAN 20    20   120   3      -    P    Master   192.168.20.2    192.168.20.254
```

从show命令的输出结果可以看到，SWB路由器在VRRP组10中，优先级为100，状态为Backup路由器；在VRRP组20中，优先级为120，状态为Master路由器。

小结

通过VRRP技术的学习，主要掌握VRRP单备份组、VRRP多备份组和基于SVI的VRRP备份组等配置与管理工作。

5.4 VPN技术

问题描述

由于主校区与分校区之间需要传输业务数据，为了保障业务数据的安全，两地之间使用专用链路来传输业务数据。为了实现业务数据的高可用性，需要使用互联网作为业务数据传输的备份链路。为了保障业务数据在互联网传输的安全性，使用IPSec VPN技术对数据进行加密，当专用链路断开之后，VPN链路启动，并加密传输业务数据。

为了方便学校教师远程办公，访问内部网络资源，并保障其安全性，需要部署远程接入VPN技术，在路由器的外部接口配置远程接入VPN，允许学校教师采用PPTP访问内网资源，假设只允许用户teacher1、teacher2、teacher3、teacher4和teacher5访问内部网络，其密码为用户名。

问题分析

配置VPN主要涉及IPSec VPN配置、Gre VPN配置、Gre over IPSec VPN PPTP VPN和

L2TP VPN配置。

1. IPSec VPN配置

步骤1 配置默认生命周期（可选）。若无特别说明，IKE将采用该生命周期值进行协商，从而使IPSec的生命周期不会超过默认生命周期的长度。参数seconds表示使用秒数。

router(config)#crypto ipsec security-association lifetime seconds *seconds*

步骤2 创建加密访问列表。使用permit关键字将使得满足指定条件的所有IP通信都受到相应加密映射条目中所描述策略的加密保护；使用deny关键字可以防止通信受到特定加密映射条目的加密保护。

router(config)#**access-list** list-number {**permit** | **deny**} protocol source source-wildcard destination destination-wildcard

步骤3 定义变换集合。transform参数是系统所支持的算法，算法可以进行一定规则的组合。

router(config)#**crypto ipsec transform-set** transform-set-name transform1 [transform2 [transform3]]
router(cfg-crypto-trans)#**mode** {**tunnel** | **transport**} （可选）

步骤4 创建加密映射条目（只讨论使用IKE来建立安全联盟情况）。

router(config)#**crypto map** *map-name seq-num* **ipsec-isakmp**
#指定要创建或修改的加密映射条目
router(config-crypto-map)#**match address** *access-list-id*
#为加密映射指定一个访问列表
router(config-crypto-map)#**set peer** {*hostname* | *ip-address*}
#指定远端IPSec对等体
router(config-crypto-map)#**set transform-set** transform-set-name1[transform-set-name2…transform-set-name6]
#指定使用哪个变换集合，按一定优先级列出变化集

步骤5 将加密映射条目应用到接口上。

router(config-if)#**crypto map** *map-name*

2. IKE配置

步骤1 开放或者关闭IKE。

router(config)#crypto isakmp enable
router(config)#no crypto isakmp enable

步骤2 创建IKE策略。

router(config)#**crypto isakmp policy** *priority*
#标识要创建的策略，每条策略由优先级唯一标识
router(config-isakmp)#encryption des | 3des
#指定加密算法
router(config-isakmp)#**hash** {**sha** | **md5**}
#指定HASH方法
router(config-isakmp)#authentication {pre-share | rsa-sig}
#指定验证方法
router(config-isakmp)#**group** {**1** | **2**}
#指定Diffie-Hellman组标识
router(config-isakmp)#**lifetime** *seconds*
#指定IKE安全联盟的生命周期

步骤3 选择工作模式。Main表示主模式，aggressive表示积极模式，

router(config)#**crypto map** *map-name seq-num* **ipsec-isakmp**

router(config-crypto-map)#set exchange-mode {main | aggressive}
#选择IKE协商的工作模式，默认情况下是主模式
router(config)#crypto ipsec-isakmp mode-detect
#根据发起方所采用的模式，采用相应的模式进行协商。作为响应方，默认情况下采用主模式来进行协商。

步骤4 配置预共享密钥。

router(config)#**ip host** *hostname address*
#如果采用hostname来标识对端身份，指定该hostname所对应的IP。
router(config)#**crypto isakmp key 0|7** *keystring* {**hostname** *peer-hostname* | **address** *peer-address*}
#指定与特定远程IKE对等体使用的共享密钥。数字0为输入明文，数字7为输入密文。
router(config)#**crypto isakmp key 0|7** *keystring* **address** *peer-address* [*mask*]
#指定针对某个网段的IKE对等体使用的共享密钥。peer-address和mask都是0.0.0.0，是默认的预共享密码。

3. Gre VPN配置

步骤1 进入指定tunnel接口配置模式，设置隧道接口地址。tunnel-number是建立的隧道接口。

router(config)#**interface tunnel** *tunnel-number*
router(config-if)#**ip address** *ip-address mask*

步骤2 配置隧道的源接口或源地址。

router(config-if)#**tunnel source** {ip-address | interface-name interface-number}

步骤3 配置隧道的目的地址。

router(config-if)#**tunnel destination** {*ip-address*}

4. PPTP/L2TP VPN服务端配置

步骤1 配置PPTP拨号地址池。Pool-name是local pool的名称，Low-ip-address表示地址池中第一个分配的IP地址，High-ip-address表示地址池中最后结束的IP地址。

router(config)# **ip local pool** {*Pool-name*} {*Low-ip-address*}{ *High-ip-address* }

步骤2 配置用户信息。username是远程拨入用户名字，password是用户密码。

router(config)#**username** *username* **password** *password*

步骤3 开启VPN拨入。

router(config)#**vpdn enable**

步骤4 创建VPN拨号组。vpdn-group-name是拨号组的名称，accept-dialin表示允许接收远程客户端拨入，protocol 表示选用哪种协议进行拨号，Virtual-Template-number设置使用的虚拟模板。

router(config)#**vpdn-group** {*vpdn-group-name*}
router(config-vpdn)#**accept-dialin**
router(config-vpdn-acc-in)#**protocol** {**any** | **pptp** | **l2tp**}
router(config-vpdn-acc-in)#**virtual-template** {*Virtual-Template-number*}

步骤5 创建虚拟拨入接口。Virtual-template-number是指定的virtual-template接口的序号，chap和pap 表示ppp验证的协议方式，ip unnumbered表示拨入地址，Pool-name是指地址池的名称。

router(config)#**interface virtual-template** {*Virtual-template-number*}
router(config-if-Virtual-Template 1)#**ppp authentication** {**chap** | **pap**}
router(config-if-Virtual-Template 1)#**ip unnumbered** {*interface*}
router(config-if-Virtual-Template 1)#**peer default ip address pool** {*Pool-name*}

任 务 单

1	IPSec site to site with preshare-key
2	Gre tunnel
3	IPSec over Gre tunnels with preshare-key
4	PPTP
5	L2TP

解决步骤

根据任务单的安排完成任务。

任务1：IPSec site to site with preshare-key

任务实施

1. 任务描述及网络拓扑设计

RT1为主校区出口路由器，RT3为分校区出口路由器。中间的ISP路由器模拟广域网。主校区和分校区的内网通过IPSec VPN相互通信。绘制拓扑结构图，如图5-14所示。

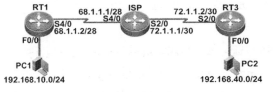

图5-14 IPSec VPN 配置

2. 网络设备配置

（1）RT1路由器配置

1）配置RT1路由器的IP地址。

RT1(config)#interface fastEthernet 0/0
RT1(config-if)#ip address 192.168.10.1 255.255.255.0
RT1(config-if)#no shutdown
RT1(config-if)#interface serial 4/0
RT1(config-if)#ip address 68.1.1.2 255.255.255.240
RT1(config)#no shutdown

2）在RT1上配置静态路由。

RT1(config)#ip route 0.0.0.0 0.0.0.0 68.1.1.1   #配置访问外网默认路由

3）在RT1路由器上配置IKE阶段一需要的使用策略。

RT1(config)#crypto isakmp enable   #启用ISAKMP

4）RT1路由器上配置预共享密钥，在两台对等体路由器上密钥必须一致。

RT1(config)#crypto isakmp key 0 aaaaaa address 72.1.1.2   #配置加密密钥

5）在RT1路由器上为IKE阶段一的协商，配置ISAKMP的策略。

RT1(config)#crypto isakmp policy 1   #配置第一阶段加密策略

```
RT1(config-isakmp)#hash md5                              #配置散列算法为md5
RT1(config-isakmp)#encryption des                        #指定加密算法为DES
RT1(config-isakmp)#authentication pre-share              #使用预共享密钥进行认证
RT1(config-isakmp)#lifetime 86400                        #指出协商后的SA的寿命
RT1(config-isakmp)#group 1                               #使用DH组1进行密钥交换
RT1(config-isakmp)#exit
```
6）在RT1路由器上配置IPsec变换集，其用于IKE阶段二的IPsec的SA协商。
```
RT1(config)#crypto ipsec transform-set my_trans esp-des  #配置第二阶段加密
RT1(cfg-crypto-trans)#mode tunnel                        #选择隧道模式
RT1(cfg-crypto-trans)#exit
```
7）在RT1路由器上配置加密访问控制列表，用于指出哪些数据流是需要加密的。
```
RT1(config)#access-list 100 permit ip 192.168.10.0 0.0.0.255 192.168.40.0 0.0.0.255
```
8）在RT1路由器上配置加密映射表，用于关联相关的变换集。
```
RT1(config)#crypto map vpn_to_RT3 10 ipsec-isakmp        #配置多个MAP条目
RT1(config-crypto-map)#set peer 72.1.1.2                 #指定对等体IP
RT1(config-crypto-map)#set transform-set my_trans        #引用IPsec的变换集
RT1(config-crypto-map)#match address 100                 #对指定的数据流进行保护
RT1(config-crypto-map)#exit
RT1(config)#exit
```
9）在RT1路由器上将加密映射表应用到需要建立的隧道接口。
```
RT1(config)#interface serial 4/0
RT1(config-if)#crypto map vpn_to_RT3                     #应用加密映射
```
（2）RT3路由器配置

1）配置RT3路由器的IP地址。
```
RT3(config)#interface fastEthernet 0/0
RT3(config-if)#ip address 192.168.40.1 255.255.255.0
RT3(config-if)#no shutdown
RT3(config-if)#interface serial 2/0
RT3(config-if)#ip address 72.1.1.2 255.255.255.252
RT3(config)#no shutdown
```
2）在RT3上配置静态路由。
```
RT3(config)#ip route 0.0.0.0 0.0.0.0 72.1.1.1
```
3）在RT3路由器上配置IKE阶段一需要使用的策略。
```
RT3(config)#crypto isakmp enable
```
4）在RT3路由器上配置预共享密钥，在两台对等体路由器上密钥必须一致。
```
RT3(config)#crypto isakmp key 0 aaaaaa address 68.1.1.2
```
5）在RT3路由器上为IKE阶段一的协商，配置ISAKMP的策略。
```
RT3(config)#crypto isakmp policy 2
RT3(config-isakmp)#hash md5
RT3(config-isakmp)#encryption des
RT3(config-isakmp)#authentication pre-share
RT3(config-isakmp)#lifetime 86400
RT3(config-isakmp)#group 1
RT3(config-isakmp)#exit
```
6）在RT3路由器上配置IPsec变换集，其用于IKE阶段二的IPsec的SA协商。
```
RT3(config)#crypto ipsec transform-set my_trans esp-des
RT3(cfg-crypto-trans)#mode tunnel
```

7）在RT3路由器上配置加密访问控制列表，用于指出哪些数据流是需要加密的。

RT3(config)#access-list 100 permit ip 192.168.40.0 0.0.0.255 192.168.10.0 0.0.0.255

8）在RT3路由器上配置加密映射表，用于关联相关的变换集。

RT3(config)#crypto map vpn_to_RT1 10 ipsec-isakmp
RT3(config-crypto-map)#set peer 68.1.1.2
RT3(config-crypto-map)#set transform-set my_trans
RT3(config-crypto-map)#match address 100

9）在RT3路由器上将加密映射表应用到需要建立的隧道接口。

RT3(config)#interface serial 2/0
RT3(config-if)#crypto map vpn_to_RT1   #应用加密映射

（3）ISP路由器配置

ISP#configure terminal
ISP(config)#interface serial 4/0
ISP(config-if)#ip address 68.1.1.1 255.255.255.240
ISP(config-if)#no shutdown
ISP(config-if)#interface serial 2/0
ISP(config-if)#ip address 72.1.1.1 255.255.255.252
ISP(config)#no shutdown

（4）测试验证

1）按拓扑图，给PC1主机配置相应的IP地址为192.168.10.2/24，网关为192.168.10.1；给PC2主机配置相应的IP地址为192.168.40.2/24，网关为192.168.40.1，PC1 ping PC2，结果如下所示。

C:\Documents and Settings\Administrator>ping 192.168.40.2
Pinging 192.168.40.2 with 32 bytes of data:
Request timed out.
Reply from 192.168.40.2: bytes=32 time<1ms TTL=63
Reply from 192.168.40.2: bytes=32 time<1ms TTL=63
Reply from 192.168.40.2: bytes=32 time<1ms TTL=63
Ping statistics for 192.168.40.2:
Packets: Sent = 4, Received = 2, Lost = 1(25% loss)

2）在RT1路由器上查看本地的IKE阶段一的安全关联。

RT1#show crypto isakmp sa
IPv4 Crypto ISAKMP SA
dst   src   state   conn-id   slot   status
72.1.1.2   68.1.1.2   QM_IDLE   1001   0   ACTIVE

3）在RT1路由器上查看IKE阶段二的IPsec的安全关联。

RT1#show crypto ipsec sa
interface: Serial4/0
    Crypto map tag: vpn_to_R3, local addr 68.1.1.2
....
    local crypto endpt.: 68.1.1.2, remote crypto endpt.: 72.1.1.2
    path mtu 1500, ip mtu 1500, ip mtu idb Serial1/1
    current outbound spi: 0xA9133A18(2836609560)

    inbound esp sas:
     spi: 0x702868C8(1881696456)
       transform: esp-des ,
       in use settings ={Tunnel, }

第5章 网络服务应用

```
        conn id: 1, flow_id: 1, crypto map: vpn_to_R3
        sa timing: remaining key lifetime (k/sec): (4436970/1326)
        IV size: 8 bytes
        replay detection support: N
        Status: ACTIVE
……………
     outbound esp sas:
       spi: 0xA9133A18(2836609560)
         transform: esp-des ,
         in use settings ={Tunnel, }
         conn id: 2, flow_id: 2, crypto map: vpn_to_R3
         sa timing: remaining key lifetime (k/sec): (4436970/1325)
         IV size: 8 bytes
         replay detection support: N
         Status: ACTIVE
……………
```

任务2：Gre tunnel

任务实施

1. 任务描述及网络拓扑设计

RT1为主校区出口路由器，RT3为分校区出口路由器。中间的ISP路由器模拟广域网。主校区和分校区的内网通过GRE隧道相互通信。绘制拓扑结构图，如图5-15所示。

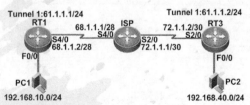

图5-15 Gre VPN 配置

2. 网络设备配置

（1）RT1路由器配置

1）配置RT1路由器的IP地址，并且使用Ping命令确认各路由器的直连口的互通。

```
RT1(config)#interface fastEthernet 0/0
RT1(config-if)#ip address 192.168.10.1 255.255.255.0
RT1(config-if)#no shutdown
RT1(config-if)#interface serial 4/0
RT1(config-if)#ip address 68.1.1.2 255.255.255.240
RT1(config)#no shutdown
```

2）在RT1上配置静态路由。

```
RT1(config)#ip route 0.0.0.0 0.0.0.0 68.1.1.1
```

3）在RT1路由器上配置隧道接口。

```
RT1(config)#interface tunnel 1                              #建立隧道接口
RT1(config-if)#ip address 61.1.1.1 255.255.255.0            #指定隧道接口地址
RT1(config-if)#tunnel destination 72.1.1.2                  #指定隧道目的地址
RT1(config-if)#tunnel source serial 4/0                     #隧道源地址为s4/0接口地址
```

RT1(config-if)#no shutdown
RT1(config-if)#exit
4）在RT1配置静态路由让内网进行通信。
RT1(config)#ip route 192.168.40.0 255.255.255.0 61.1.1.2
（2）RT3路由器配置
1）配置RT3路由器的IP地址。
RT3(config)#interface fastEthernet 0/0
RT3(config-if)#ip address 192.168.40.1 255.255.255.0
RT3(config-if)#no shutdown
RT3(config-if)#interface serial 2/0
RT3(config-if)#ip address 72.1.1.2 255.255.255.252
RT3(config)#no shutdown
2）在RT3上配置静态路由。
RT3(config)#ip route 0.0.0.0 0.0.0.0 72.1.1.1
3）在RT1路由器上配置隧道接口。

RT3(config)#interface tunnel 1   #建立隧道接口
RT3(config-if)#ip add 61.1.1.2 255.255.255.0   #指定隧道接口地址
RT3(config-if)#tunnel destination 68.1.1.2   #指定隧道目的地址
RT3(config-if)#tunnel source serial 2/0   #隧道源地址为s2/0接口地址
RT3(config-if)#no shutdown

4）在RT3上配置静态路由让内网进行通信。
RT3(config)#ip route 192.168.10.0 255.255.255.0 61.1.1.1
（3）ISP路由器配置
ISP(config)#interface serial 2/0
ISP(config-if)#ip address 72.1.1.1 255.255.255.252
ISP(config-if)#no shutdown
ISP(config-if)#exit
ISP(config)#interface serial 4/0
ISP(config-if)#ip address 68.1.1.1 255.255.255.240
ISP(config-if)#no shutdown
（4）测试网络连通性
　　按拓扑图，给PC1主机配置的相应IP地址为192.168.10.2/24，网关为192.168.10.1；给PC2主机配置的相应IP地址为192.168.40.2/24，网关为192.168.40.1，PC1 ping PC2，结果如下所示。
C:\Documents and Settings\Administrator>ping 192.168.40.2
Pinging 192.168.40.2 with 32 bytes of data:
Reply from 192.168.40.2: bytes=32 time<1ms TTL=254
Reply from 192.168.40.2: bytes=32 time<1ms TTL=254
Reply from 192.168.40.2: bytes=32 time<1ms TTL=254
Reply from 192.168.40.2: bytes=32 time<1ms TTL=254
Ping statistics for 192.168.40.2:
Packets: Sent = 4, Received = 4, Lost = 0(0% loss)

任务3：Gre over IPSec tunnels with preshare-key

任务实施

1. 任务描述及网络拓扑设计

　　RT1为主校区出口路由器，RT3为分校区出口路由器。中间的ISP路由器模拟广域网。

第5章 网络服务应用

主校区和分校区的内网通过GRE隧道相互通信,通过IPSEC将GRE的流量加密,从而起到保护内网数据流的作用。绘制拓扑结构图,如图5-16所示。

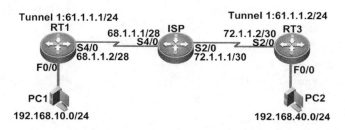

图5-16　IPSec over Gre VPN配置

2. 网络设备配置

（1）RT1路由器配置

1）配置RT1路由器的IP地址,并且使用Ping命令确认各路由器的直连口的互通。

RT1(config)#interface fastEthernet 0/0
RT1(config-if)#ip address 192.168.10.1 255.255.255.0
RT1(config-if)#no shutdown
RT1(config-if)#exit
RT1(config)#interface serial 4/0
RT1(config-if)#ip address 68.1.1.2 255.255.255.240
RT1(config)#no shutdown

2）在RT1路由器上配置隧道接口。

RT1(config)#interface tunnel 1
RT1(config-if)#ip address 61.1.1.1 255.255.255.0
RT1(config-if)#tunnel destination 72.1.1.2
RT1(config-if)#tunnel source serial 4/0
RT1(config-if)#no shutdown

3）在RT1上配置静态路由。

RT1(config)#ip route 192.168.40.0 255.255.255.0 61.1.1.2
RT1(config)#ip route 0.0.0.0 0.0.0.0 68.1.1.1

4）在RT1路由器上配置预共享密钥,在两台对等体路由器上密钥必须一致。

RT1(config)#crypto isakmp key 0 ciscokey address 72.1.1.2　　　　#配置加密密钥

5）在RT1路由器上配置IKE阶段一需要的使用策略。

RT1(config)#crypto isakmp policy 1　　　　　　　　　　#配置第一阶段加密策略
RT1(config-isakmp)#hash md5　　　　　　　　　　　　　#配置散列算法为md5
RT1(config-isakmp)#encryption des　　　　　　　　　　　#指定加密算法为DES
RT1(config-isakmp)#authentication pre-share　　　　　　　#使用预共享密钥进行认证
RT1(config-isakmp)#lifetime 86400　　　　　　　　　　　#指出协商后的SA的寿命
RT1(config-isakmp)#group 1　　　　　　　　　　　　　　#使用DH组1进行密钥交换

6）在RT1路由器上配置IPsec变换集,其用于IKE阶段二的IPsec的SA协商。

RT1(config)#crypto ipsec transform-set my_trans esp-des　　　#配置第二阶段加密策略
RT1(cfg-crypto-trans)#mode transport　　　　　　　　　　　#隧道模式为传输模式

7）在RT1路由器上配置加密访问控制列表,用于加密GRE隧道的数据流。

RT1(config)#access-list 100 permit gre host 68.1.1.2 host 72.1.1.2

8）在RT1路由器上配置加密映射表,用于关联相关的变换集。

```
RT1(config)#crypto map gre_to_RT3 10 ipsec-isakmp        #配置多个MAP条目
RT1(config-crypto-map)#set peer 72.1.1.2                  #指定对等体IP
RT1(config-crypto-map)#set transform-set my_trans         #引用IPsec的变换集
RT1(config-crypto-map)#match address 100                  #对GRE数据流进行保护
RT1(config-crypto-map)#exit
```

9）在RT1路由器上将加密映射表应用到需要建立的隧道接口。

```
RT1(config)#interface s4/0
RT1(config-if)#crypto map gre_to_RT3                      #应用加密映射
```

（2）RT3路由器配置

1）配置RT3路由器的IP地址。

```
RT3(config)#interface fastEthernet 0/0
RT3(config-if)#ip address 192.168.40.1 255.255.255.0
RT3(config-if)#no shutdown
RT3(config-if)#exit
RT3(config)#interface serial 2/0
RT3(config-if)#ip address 72.1.1.2 255.255.255.252
RT3(config)#no shutdown
```

2）在RT3路由器上配置隧道接口。

```
RT3(config)#interface tunnel 1
RT3(config-if)#ip address 61.1.1.2 255.255.255.0
RT3(config-if)#tunnel destination 68.1.1.2
RT3(config-if)#tunnel source serial 2/0
RT3(config-if)#no shutdown
```

3）在RT3上配置静态路由。

```
RT3(config)#ip route 0.0.0.0 0.0.0.0 72.1.1.1
RT3(config)#ip route 192.168.10.0 255.255.255.0 61.1.1.1
```

4）在RT3路由器上配置预共享密钥，在两台对等体路由器上密钥必须一致。

```
RT3(config)#crypto isakmp enable
RT3(config)#crypto isakmp key 0 ciscokey address 68.1.1.2
```

5）在RT3路由器上配置IKE阶段一需要的使用策略。

```
RT3(config)#crypto isakmp policy 1
RT3(config-isakmp)#hash md5
RT3(config-isakmp)#encryption des
RT3(config-isakmp)#authentication pre-share
RT3(config-isakmp)#group 1
RT3(config-isakmp)#exit
```

6）在RT3路由器上配置IPsec变换集，其用于IKE阶段二的IPsec的SA协商。

```
RT3(config)#crypto ipsec transform-set my_trans esp-des
RT3(cfg-crypto-trans)#mode transport
```

7）在RT3路由器上配置加密访问控制列表，用于加密GRE隧道的数据流。

```
RT3(config)# access-list 100 permit gre host 72.1.1.2 host 68.1.1.2
```

8）在RT3路由器上配置加密映射表，用于关联相关的变换集。

```
RT3(config)#crypto map gre_to_RT1 10 ipsec-isakmp
RT3(config-crypto-map)#set peer 72.1.1.2
RT3(config-crypto-map)#set transform-set my_trans
RT3(config-crypto-map)#match address 100
```

9）在RT3路由器上将加密映射表应用到需要建立的隧道接口。

```
RT3(config)#interface s2/0
RT3(config-if)#crypto map gre_to_RT1
```
（3）ISP路由器配置
```
ISP(config)# interface serial 2/0
ISP(config-if)#ip address 72.1.1.1 255.255.255.252
ISP(config-if)#no shutdown
ISP(config-if)#exit
ISP(config-if)#interface serial 4/0
ISP(config-if)#ip address 68.1.1.1 255.255.255.240
ISP(config)#no shutdown
```
（4）测试验证

1）按拓扑图，给PC1主机配置相应的IP地址为192.168.10.2/24，网关为192.168.10.1；给PC2主机配置相应的IP地址为192.168.40.2/24，网关为192.168.40.1，PC1 ping PC2，结果如下所示。

```
C:\Documents and Settings\Administrator>ping 192.168.40.2
Pinging 192.168.40.2 with 32 bytes of data:

Reply from 192.168.40.2: bytes=32 time<1ms TTL=63
Reply from 192.168.40.2: bytes=32 time<1ms TTL=63
Reply from 192.168.40.2: bytes=32 time<1ms TTL=63
Reply from 192.168.40.2: bytes=32 time<1ms TTL=63
Ping statistics for 192.168.40.2:
Packets: Sent = 4, Received = 4, Lost = 0(0% loss)
```
2）在RT1路由器上查看本地的IKE阶段一的安全关联。
```
RT1#show crypto isakmp sa
IPv4 Crypto ISAKMP SA
dst              src              state          conn-id    slot    status
72.1.1.2         68.1.1.2         QM_IDLE        1001       0       ACTIVE
```
#出现ACTIVE 表示协商成功

3）在RT1路由器上查看IKE阶段二的IPsec的安全关联。
```
RT1#show crypto ipsec sa
interface: Serial 4/0
    Crypto map tag: gre_to_RT3, local addr 210.26.190.1
   protected vrf: (none)
   local  ident (addr/mask/prot/port): (68.1.1.2/255.255.255.255/47/0)
   remote ident (addr/mask/prot/port): (72.1.1.2/255.255.255.255/47/0)
   current_peer 72.1.1.2 port 500
     PERMIT, flags={origin_is_acl,}
    #pkts encaps: 19, #pkts encrypt: 19, #pkts digest: 19 //有多个数据包被加解密
    #pkts decaps: 5, #pkts decrypt: 5, #pkts verify: 5
    #pkts compressed: 0, #pkts decompressed: 0
    #pkts not compressed: 0, #pkts compr. failed: 0
    #pkts not decompressed: 0, #pkts decompress failed: 0
    #send errors 1, #recv errors 0

     local crypto endpt.: 72.1.1.2, remote crypto endpt.: 68.1.1.2
     path mtu 1500, ip mtu 1500, ip mtu idb Serial 1/1
     current outbound spi: 0x1175F98F(292944271)
```

....
　　　　inbound esp sas:
　　　　 spi: 0xCB8175ED(3414259181)
　　　　　　transform: esp-des ,
　　　　　　in use settings ={Transport, }
　　　　　　conn id: 1, flow_id: 1, crypto map: gre_to_RT3
　　　　　　sa timing: remaining key lifetime (k/sec): (4527485/3084)
　　　　　　IV size: 8 bytes
　　　　　　replay detection support: N
　　　　　　Status: ACTIVE

　　　　outbound esp sas:
　　　　 spi: 0x1175F98F(292944271)
　　　　　　transform: esp-des ,
　　　　　　in use settings ={Transport, }
　　　　　　conn id: 2, flow_id: 2, crypto map: gre_to_RT3
　　　　　　sa timing: remaining key lifetime (k/sec): (4527483/3084)
　　　　　　IV size: 8 bytes
　　　　　　replay detection support: N
　　　　　　Status: ACTIVE

任务4：PPTP

任务实施

1. 任务描述及网络拓扑设计

为了方便学校教师远程办公，访问内部网络资源，并保障其安全性，需要部署远程接入VPN技术，在路由器的外部接口配置远程接入VPN，允许学校教师采用PPTP访问内网资源，假设只允许用户teacher1、teacher2、teacher3、teacher4和teacher5访问内部网络，其密码为用户名，其获取的IP地址为192.168.30.150～192.168.30.200。绘制拓扑结构图，如图5-17所示。

图5-17　PPTP配置

2. 网络设备配置

（1）RT1路由器配置

1）配置RT1路由器的IP地址。

RT1(config)#interface fastEthernet 0/0
RT1(config-if)#ip address 192.168.10.1 255.255.255.0
RT1(config-if)#no shutdown
RT1(config-if)#exit

```
RT1(config)#interface serial 4/0
RT1(config-if)#ip address 68.1.1.2 255.255.255.240
RT1(config)#no shutdown
```
2）在RT1路由器上配置PPTP。

```
RT1(config)#username teacher1 password teacher1        #配置用户信息
RT1(config)#username teacher2 password teacher2        #配置用户信息
RT1(config)#username teacher3 password teacher3        #配置用户信息
RT1(config)#username teacher4 password teacher4        #配置用户信息
RT1(config)#username teacher5 password teacher5        #配置用户信息
RT1(config)#ip local pool vpn 192.168.30.150 192.168.30.200    #配置拨号地址池
RT1(config)#vpdn enable                                #开启VPN拨入
RT1(config-vpdn)#vpdn-group pptp                       #设置拨号组的名称
RT1(config-vpdn)#Default PPTP VPDN group
RT1(config-vpdn)#accept-dialin                         #允许远程客户端拨入
RT1(config-vpdn-acc-in)#protocol pptp                  #选用哪种协议进行拨号
RT1(config-vpdn-acc-in)#virtual-template 1             #设置使用的虚拟模板
RT1(config-vpdn-acc-in)#exit
RT1(config)#interface Virtual-Template 1               #定义虚拟模板1
RT1(config-if)#ppp authentication chap                 #启用PPP认证
RT1(config-if)#ip unnumbered FastEthernet 0/1          #指定拨入地址
RT1(config-if)#peer default ip address pool vpn        #设置对端IP地址
RT1(config-if)#exit
```
3）在RT1上配置静态路由，确保主校区和分校区可以相互通信。

```
RT1(config)#ip route 0.0.0.0 0.0.0.0 68.1.1.1
```
（2）RT3路由器配置

1）配置RT3路由器的IP地址。

```
RT3(config)#interface fastEthernet 0/0
RT3(config-if)#ip address 192.168.40.2 255.255.255.0
RT3(config-if)#no shutdown
RT3(config-if)#exit
RT3(config)#interface serial 2/0
RT3(config-if)#ip address 72.1.1.2 255.255.255.252
RT3(config)#no shutdown
```
2）在RT3上配置静态路由，确保主校区和分校区可以相互通信。

```
RT3(config)#ip route 0.0.0.0 0.0.0.0 72.1.1.1
```
（3）ISP路由器配置

```
ISP(config)# interface serial 2/0
ISP(config-if)#ip address 72.1.1.1 255.255.255.252
ISP(config-if)#no shutdown
ISP(config-if)#exit
ISP(config)#interface serial 4/0
ISP(config-if)#ip address 68.1.1.1 255.255.255.240
ISP(config)#no shutdown
```
（4）测试验证

1）在教师远程办公主机上创建PPTP连接。

步骤1　在"网上邻居"上打开"属性"对话框，单击"创建一个新的连接"，如图5-18所示。

图5-18 创建新连接

步骤2 单击"下一步"按钮，如图5-19所示。

图5-19 单击"下一步"按钮

步骤3 选择"连接到我的工作场所的网络"，如图5-20所示，单击"下一步"按钮。

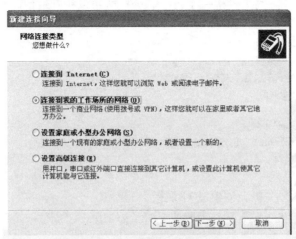

图5-20 选择网络连接类型

第5章 网络服务应用

步骤4 选择"虚拟专用网络连接",如图5-21所示,单击"下一步"按钮。

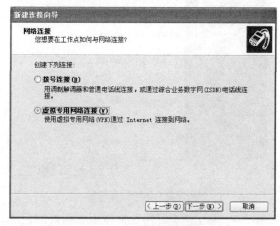

图5-21 选择网络连接

步骤5 输入公司名,单击"下一步"按钮,如图5-22所示。

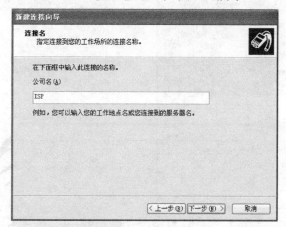

图5-22 输入公司名

步骤6 输入VPN服务器IP地址,单击"下一步"按钮,如图5-23所示。

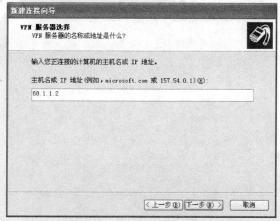

图5-23 输入VPN服务器IP地址

步骤7 单击"完成"按钮,完成创建连接。如图5-24所示。

图5-24 完成创建连接

步骤8 单击"连接isp"对话框的"属性"按钮,如图5-25所示。在打开的如图5-26所示的对话框中选择"高级"安全选项的单选按钮,单击"设置"按钮。

图5-25 "连接isp"对话框　　　　　图5-26 "isp属性"对话框

步骤9 在"数据加密"方式的下拉列表中选择"可选加密",勾选"质询握手身份验证协议(CHAP)"复选框,如图5-27所示。

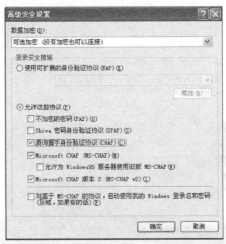

图5-27 "高级安全设置"对话框

第5章 网络服务应用

步骤10 单击"确定"按钮后拨入。

2)按拓扑图,给PC1主机配置的相应IP地址为192.168.10.2/24,网关为192.168.10.1,给PC2主机配置的相应IP地址为192.168.40.2/24,网关为192.168.40.1,教师远程办公主机的IP地址为100.1.1.2/30,网关为100.1.1.1。用教师远程办公主机ping PC1,结果如下所示。

```
C:\Documents and Settings\Administrator>ping 192.168.10.2
Pinging 192.168.10.2with 32 bytes of data:
Reply from 192.168.10.2: bytes=32 time<1ms TTL=254
Reply from 192.168.10.2: bytes=32 time<1ms TTL=254
Reply from 192.168.10.2: bytes=32 time<1ms TTL=254
Reply from 192.168.10.2: bytes=32 time<1ms TTL=254
Ping statistics for 192.168.10.2:
Packets: Sent = 4, Received = 4, Lost = 0(0% loss)
```

任务5:L2TP

任务实施

1. 任务描述及网络拓扑设计

为了方便学校教师远程办公,访问内部网络资源,并保障其安全性,需要部署远程接入VPN技术,在路由器的外部接口配置远程接入VPN,允许学校教师采用L2TP访问内网资源,假设只允许用户teacher1、teacher2、teacher3、teacher4和teacher5访问内部网络,其密码为用户名,其获取的IP地址为192.168.30.150~192.168.30.200。绘制拓扑结构图,如图5-28所示。

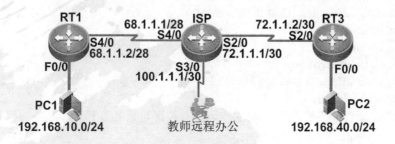

图5-28 L2TP配置

2. 网络设备配置

(1)RT1路由器配置

1)配置RT1路由器的IP地址。

```
RT1(config)#interface fastEthernet 0/0
RT1(config-if)#ip address 192.168.10.1 255.255.255.0
RT1(config-if)#no shutdown
RT1(config-if)#exit
RT1(config)#interface serial 4/0
RT1(config-if)#ip address 68.1.1.2 255.255.255.240
```

RT1(config)#no shutdown

2）在RT1路由器配置L2TP。

RT1(config)#username teacher1 password teacher1
RT1(config)#username teacher2 password teacher2
RT1(config)#username teacher3 password teacher3
RT1(config)#username teacher4 password teacher4
RT1(config)#username teacher5 password teacher5
RT1(config)#ip local pool vpn 192.168.30.150 192.168.30.200
RT1(config)#vpdn enable
RT1(config-vpdn)#vpdn-group l2tp
RT1(config-vpdn)#Default l2tp VPDN group
RT1(config-vpdn)#accept-dialin
RT1(config-vpdn-acc-in)#protocol l2tp
RT1(config-vpdn-acc-in)#virtual-template 1
RT1(config-vpdn-acc-in)#exit
RT1(config)#interface Virtual-Template 1
RT1(config-if)#ppp authentication chap
RT1(config-if)#ip unnumbered FastEthernet 0/1
RT1(config-if)#peer default ip address pool vpn
RT1(config-if)#exit

3）在RT1上配置静态路由。确保主校区和分校区可以相互通信。

RT1(config)#ip route 0.0.0.0 0.0.0.0 68.1.1.1

（2）RT3路由器配置

1）配置RT3路由器的IP地址。

RT3(config)#interface fastEthernet 0/0
RT3(config-if)#ip address 192.168.40.2 255.255.255.0
RT3(config-if)#no shutdown
RT3(config-if)#exit
RT3(config)#interface serial 2/0
RT3(config-if)#ip address 72.1.1.2 255.255.255.252
RT3(config)#no shutdown

2）在RT3上配置静态路由。确保主校区和分校区可以相互通信。

RT3(config)#ip route 0.0.0.0 0.0.0.0 72.1.1.1

（3）ISP路由器配置

ISP(config)#interface serial 2/0
ISP(config-if)#ip address 72.1.1.1 255.255.255.252
ISP(config-if)#no shutdown
ISP(config-if)#exit
ISP(config)#interface serial 4/0
ISP(config-if)#ip address 68.1.1.1 255.255.255.240
ISP(config)#no shutdown

（4）测试验证

1）在教师办公主机上创建L2TP连接。

步骤1 在"网上邻居"上打开"属性"对话框，单击"创建一个新的连接"，如图5-29所示。

第5章 网络服务应用

图5-29 创建一个新连接

步骤2 单击"下一步"按钮,如图5-30所示。

图5-30 单击"下一步"按钮

步骤3 选择"连接到我的工作场所的网络",单击"下一步"按钮,如图5-31所示。

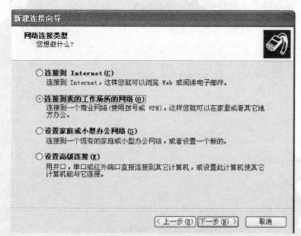

图5-31 选择网络连接类型

步骤4 选择"虚拟专用网络连接",单击"下一步"按钮,如图5-32所示。

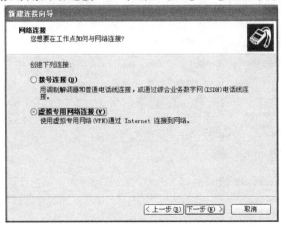

图5-32 创建网络连接

步骤5 输入公司名,单击"下一步"按钮,如图5-33所示。

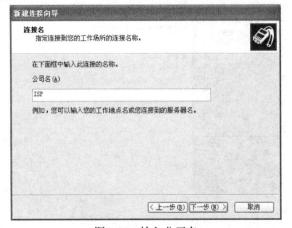

图5-33 输入公司名

步骤6 输入VPN服务器IP地址,单击"下一步"按钮,如图5-34所示。

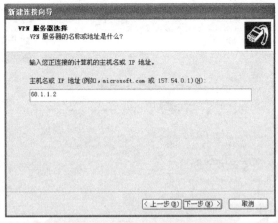

图5-34 输入VPN服务器IP地址

步骤7 单击"完成"按钮,完成创建连接,如图5-35所示。

第5章 网络服务应用

图5-35 完成创建连接

步骤8 单击"连接isp"对话框的"属性"按钮，如图5-36所示。在打开的图5-37所示的对话框中选择"高级（自定义设置）"单选按钮，单击"设置"按钮。

图5-36 打开"连接isp"对话框

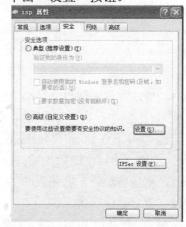

图5-37 打开"isp属性"对话框

步骤9 在"数据加密"方式的下拉列表中选择"可选加密"，勾选"质询握手身份验证协议（CHAP）"复选框，如图5-38所示。

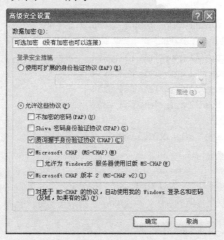

图5-38 "高级安全设置"对话框

步骤10 再单击"网络"选项卡,在"VPN类型"下拉列表中选择L2TP IPSec VPN,如图5-39所示。

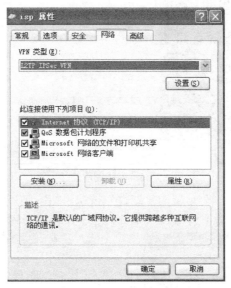

图5-39 选择VPN类型

步骤11 修改注册表键值,使L2TP忽略证书验证。

进入桌面上的"开始"→"运行"文本框,输入"Regedt32",打开"注册表编辑器",定位"HKEY_Local_Machine / System / CurrentControl Set / Services / RasMan / Parameters"主键。为该主键添加键值:ProhibitIpSec;数据类型:reg_dword;值:1。保存所做的修改,重新启动计算机使改动生效。

步骤12 重启完成后拨入。

2)按拓扑图,给PC1主机配置相应的IP地址为192.168.10.2/24,网关为192.168.10.1;给PC2主机配置相应的IP地址为192.168.40.2/24,网关为192.168.40.1,教师远程办公主机的IP地址为100.1.1.2/30,网关为100.1.1.1。用教师远程办公主机ping PC1,结果如下所示。

C:\Documents and Settings\Administrator>ping 192.168.10.2
Pinging 192.168.10.2with 32 bytes of data:
Reply from 192.168.10.2: bytes=32 time<1ms TTL=254
Reply from 192.168.10.2: bytes=32 time<1ms TTL=254
Reply from 192.168.10.2: bytes=32 time<1ms TTL=254
Reply from 192.168.10.2: bytes=32 time<1ms TTL=254
Ping statistics for 192.168.10.2:
Packets: Sent = 4, Received = 4, Lost = 0(0% loss)

小结

通过VPN技术的学习,主要掌握IPSec VPN、Gre VPN、Gre over IPsec VPN、PPTP等配置与管理工作。

VPN通过Internet传输私有用户数据流,使用加密和隧道技术来确保数据的安全,与传

第5章　网络服务应用

统的WAN连接相比，VPN在费用、灵活性、管理等方面具有优势。

配置IPSec VPN时，要注意保护流量配置对应的ACL表顺序要对应图所指定的方向。如果在接口上启用NAT，一定要将保护流量的网段从NAT中拒绝掉，因为如果感兴趣流量通过NAT会导致源地址发生变化，从而不对应ACL表，导致两边的内网不通。验证时发过去的第一个包会超时，那是因为加密算法还在协商，在协商结束后就能正常通信了。

配置Gre VPN指定Gre隧道时，不要混淆隧道源地址和隧道目的地址。

配置Gre over IPSec VPN时，IPSec配置与Site To Site 的配置相似，不同点是ACL列表写的是Gre的流量而不是内网感兴趣的流量，隧道模式改为传输模式，因为现在IPSec只是做加密，没有用到隧道功能。

在做PPTP实验时要注意模板中的非关联口一定要配置，否则会导致无法拨入或者拨入异常。

L2TP实验与PPTP实验类似，唯一不同点在于L2TP需要IPSEC支持，如果想要L2TP忽略IPSEC，必须修改注册表。

5.5　QoS技术

问题描述

随着Internet在全球的发展和社会信息化程度的提高，人们对网络的要求也越来越高，信息化需求已从单纯的数据信息向交互式多媒体信息发展，从分别服务向数据、语音、图像统一服务和网络传输发展。带宽延迟、抖动敏感且实时性强的语音、图像及其他的重要数据越来越多地在网上传输，一方面使得网络资源得到了极大的丰富，另一方面，由于数据、语音、图像等业务在延时、吞吐量或丢失率等方面有不同的要求，也就引入了如何保证网络服务质量的问题。

该校业务种类较多以及资源较为丰富，需要针对各种不同的需求，提供不同的服务质量。对实时性强且重要的数据报文提供更好的服务质量，并优先进行处理；而对于实时性不强的普通数据报文，则提供较低的处理优先级。

问题分析

若要在网络中承载各种不同的业务，就要求网络不仅能提供单一的服务，而且能为不同业务提供不同的QoS，配置QoS服务时，主要涉及自定义队列配置、优先级队列配置、基于类的加权公平队列配置和流量监管配置。

1. 自定义队列配置

步骤1　确定自定义排队的最大容量。list-number队列列表的号码，可取1～16的数字；

queue-number队列号码，可取1～16的任何数字；limit-number队列允许容纳数据包的最大个数。

  router(config)#**queue-list** *list-number* **queue** *queue-number* **limit** *limit-number*
 步骤2 将数据包分配到自定义队列。
  router(config)#**queue-list** *list-number* **protocal** *protocal-name queue-number [queue-keyword] [keyword-value]*
  #根据协议类型，来分配数据包到指定的自定义队列
  router(config)#**queue-list** *list-number* **interface** *interface-type interface-number queue-number*
  #根据数据包进入到设备接口类型，来分配数据包到指定的自定义队列
  router(config)#**queue-list** *list-number* **default** *queue-number*
  #给那些在自定义列表中不匹配任何规则的数据包分配一个自定义队列
  #参数protocal-name是协议类型，queue-keyword和keyword-value是针对各种协议的一些选项，默认的自定义队列号是1。
 步骤3 在接口上应用自定义排队队列。
  router(config-if)#**custom-queue-list** *list-number*

 2. 优先级队列配置

 步骤1 确定优先级排队方式的队列最大容量。list-number是队列列表的号码，可取1～16的任何数字；high-limit、midium-limit、normal-limit、low-limit分别代表高、中、正常、低优先级队列所能够容纳的最大数据包个数。
  router(config)#**priority-list** *list-number* **queue-limit**[*high-limit [midium-limit [normal-limit [low-limit]]]]*
 步骤2 将数据包分配到优先级队列。
  router(config)#**priority-list** *list-number* **protocal** *protocal-name* {**high**|**medium**|**normal**|**low**} [*queue-keyword] [keyword-value]*
  #根据协议类型，来分配数据包到指定的优先级队列
  router(config)#**priority-list** *list-number* **interface** *interface-type interface-number* {**high**|**medium**|**normal**|**low**}
  #根据数据包进入到设备接口类型，来分配数据包到指定的优先级队列
  router(config)#priority-list *list-number* default {high|medium|normal|low}
  #给不匹配任何规则的数据包分配到默认优先级队列
  #参数protocal-name是协议类型，queue-keyword和keyword-value是针对各种协议的一些选项；默认优先级队列为normal。
 步骤3 在接口上应用优先级排队队列。
  router(config)# **priority-group** *list-number*

 3. 基于类的加权公平队列配置

 步骤1 定义类映射表。
  router(config)#**class-map** match-all class-map-name
  #创建类映射表，要满足类映射表下的所有条件
  router(config)#**class-map** match-any class-map-name
  #创建类映射表，只要满足类映射表下的一个条件
  router(config-cmap)#**match access-group** *access-list-number*
  router(config-cmap)#**match input-interface** *interface-name*
  router(config-cmap)#**match ip dscp** *value*
  router(config-cmap)#**match ip precedence** *vlaue*
  router(config-cmap)#**match not match-type** *value*
  #设置网络数据包分类规则（按照ACL，网络数据包接口，封装协议类型，IP DSCP编码，IP

第5章 网络服务应用

Precedence编码或者以上分类规则的取非条件）。class-map-name表示类映射表名称。match-all要满足映射表的所有条件。默认创建的类映射表是match-all。match-any表示只要满足类映射表的一个条件。access-list-number表示访问控制列表号。interface-name表示网络接口名称。ip dscp表示网络数据包IP TOS域的DSCP编码值。ip precedence表示网络数据包IP TOS域precedence值。not match-type表示分类规则的取非条件。

步骤2 在规则映射表中设置类规则。

router(config)#**policy-map** *policy-map-name*
#创建规则映射表
router(config-pmap)#**class** *class-map-name*
#引用已定义的映射表
router(config-pmap)#**bandwidth** {*bandwidth-kbps* | **percent** *percent-number*}
#为指定类型的数据流分配带宽
router(config-pmap)#**queue-limit** *number-of-packets*
#设定队列深度
#参数policy-map-name表示规则映射表名称，class-map-name表示类映射表名称，bandwidth-kbps表示分配的带宽（以kbit/s为单位），number-of-packets表示CBWFQ队列深度。

步骤3 在指定接口上应用服务规则。

router(config-if)#**service-policy output** *policy-map-name*
#启用CBWFQ并指定应用的规则映射表
router(config-if)#**service-policy input** *policy-map-name*
#对入接口的报文启用policy-map策略

4. 流量监管配置

步骤1 指定要进行限速的接口。

router(config)#**interface** interface-type interface-number

步骤2 对接口的所有流量进行入接口或者出接口的报文限速。

router(config-if)#**rate-limit**{input | output} *bps burst-normal burst-max* conform-action *action* exceed-action *action*
#参数input| output 表示用户希望限制输入或者输出的流量；bps表示用户希望该流量的速率上限，单位是bit/s；burst-normal burst-max表示令牌桶（token bucket）的大小值（"令牌桶"是指网络设备的内部存储池，而"令牌"则是指以给定速率填充令牌桶的虚拟信息包。通过虚拟信息包来判断当前流进报文速率），单位是bytes；conform-action表示在速率限制以下的流量处理策略；exceed-action表示超过速率限制的流量处理策略。

任 务 单

1	custom-queue（自定义队列）
2	Priority-queue（优先级队列）
3	CBWFQ（基于类的加权公平队列）
4	Rate-limit（流量监管）

根据任务单的安排完成任务。

任务1：custom-queue（自定义队列）

任务实施

1. 任务描述及网络拓扑设计

   RT1为主校区出口路由器，SW1为主校区汇聚层交换机。在SW1上创建定制队列让VLAN10的流量优先从SW1的F0/24出去访问外网资源。绘制拓扑结构图，如图5-40所示。

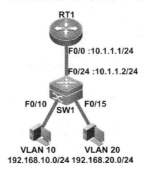

图5-40　自定义队列配置

2. 网络设备配置

（1）配置SW1交换机的IP地址

SW1(config)#vlan 10
SW1(config-vlan)#exit
SW1config)#vlan 20
SW1(config-vlan)#exit
SW1(config)#interface range fastEthernet 0/6-10
SW1(config-if-range)#switchport access vlan 10
SW1(config-if-range)#exit
SW1(config)#interface range fastEthernet 0/11-15
SW1(config-if-range)#switchport access vlan 20
SW1(config-if-range)#exit
SW1(config)#interface vlan 10
SW1(config-VLAN 10)#ip address 192.168.10.1 255.255.255.0
SW1(config-VLAN 10)#exit
SW1(config)#interface vlan 20
SW1(config-VLAN 20)#ip address 192.168.20.1 255.255.255.0
SW1(config-VLAN 20)#exit
SW1(config)#interface fastEthernet 0/24
SW1(config-if)#no switchport
SW1(config-if)#ip address 10.1.1.2 255.255.255.0

（2）在SW1上配置静态路由

SW1(config)#ip route 0.0.0.0 0.0.0.0 10.1.1.1

（3）定义流量

SW1(config)#access-list 101 permit ip 192.168.10.0 0.0.0.255 any
SW1(config)#access-list 102 permit ip 192.168.20.0 0.0.0.255 any

（4）队列排序

SW1(config)#queue-list 1 protocol ip 1 list 101　　　#将与List101匹配的流量排在第一位

第5章 网络服务应用

```
SW1(config)#queue-list 1 protocol ip 2 list 102      #将与List102匹配的流量排在第二位
SW1(config)# interface fastEthernet 0/24             #将CQ应用到接口
SW1(config-if)#custom-queue-list 1                   #将这个定制好的队列应用到接口上
```

（5）RT1路由器基本配置

```
RT1(config)#interface fastEthernet 0/0
RT1(config-if)#ip address 10.1.1.1 255.255.255.0
RT1(config-if)#no shutdown
RT1(config-if)#exit
RT1(config)#ip route 0.0.0.0 0.0.0.0 10.1.1.2
```

（6）验证测试

```
SW1#show queueing                                    #查看队列
Current fair queue configuration:
```

Interface	Discard threshold	Dynamic queues	Reserved queues	Link queues	Priority queues
BRI0	64	16	0		1
BRI0:1	64	16	0	8	1
BRI0:2	64	16	0	8	1
fastEthernet 0/24	64	256	0	8	1

```
Current DLCI priority queue configuration:
Current priority queue configuration:
Current custom queue configuration:
List  Queue   Args
1     1       protocol ip      list 101
1     2       protocol ip      list 102
```

任务2：Priority-queue（优先级队列）

任务实施

1. 任务描述及网络拓扑设计

RT1为主校区出口路由器，SW1为主校区汇聚层交换机。在SW1上创建优先级队列让HTTP通过SW1的优先级高于FTP。绘制拓扑结构图，如图5-41所示。

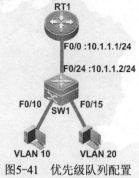

图5-41　优先级队列配置

2. 网络设备配置

（1）配置SW1交换机的IP地址

```
SW1(config)#vlan 10
SW1(config-vlan)#exit
SW1config)#vlan 20
```

SW1(config-vlan)#exit
SW1(config)#interface range fastEthernet 0/6-10
SW1(config-if-range)#switchport access vlan 10
SW1(config-if-range)#exit
SW1(config)#interface range fastEthernet 0/11-15
SW1(config-if-range)#switchport access vlan 20
SW1(config-if-range)#exit
SW1(config)#interface vlan 10
SW1(config-VLAN 10)#ip address 192.168.10.1 255.255.255.0
SW1(config-VLAN 10)#exit
SW1(config)#interface vlan 20
SW1(config-VLAN 20)#ip address 192.168.20.1 255.255.255.0
SW1(config-VLAN 20)#exit
SW1(config)#interface fastEthernet 0/24
SW1(config-if)#no switchport
SW1(config-if)#ip address 10.1.1.2 255.255.255.0

（2）在SW1上配置静态路由

SW1(config)#ip route 0.0.0.0 0.0.0.0 10.1.1.1

（3）队列排序

SW1(config)#priority-list 2 protocol ip high tcp www
            #指定WWW数据包到高优先级队列
SW1(config)#priority-list 2 protocol ip low tcp ftp
            #指定FTP数据包到低优先级队列
SW1(config)#priority-list 2 protocol ip low tcp ftp-data
            #指定FTP数据包到低优先级队列

（4）将PQ应用到接口

SW1(config)#interface fastEthernet 0/24
    SW1(config-if)#priority-group 2   #应用队列

（5）RT1的配置

RT1(config)#interface fastEthernet 0/0
RT1(config-if)#ip address 10.1.1.1 255.255.255.0
RT1(config-if)#no shutdown
RT1(config-if)#exit
RT1(config)#ip route 0.0.0.0 0.0.0.0 10.1.1.2

（6）验证测试

    SW1#show queueing   #查看队列
Current fair queue configuration:
Current priority queue configuration:
List   Queue   Args
2   high   protocol ip   tcp port www
2   low   protocol ip   tcp port ftp
2   low   protocol ip   tcp port ftp-data

任务3：CBWFQ（基于类的加权公平队列）

任务实施

1. 任务描述及网络拓扑设计

RT1为主校区出口路由器，SW1为主校区汇聚层交换机。在SW1上创建CBWFQ为HTTP

第5章 网络服务应用

分配流量带宽的5%，为FTP分配流量的25%，绘制拓扑结构图，如图5-42所示。

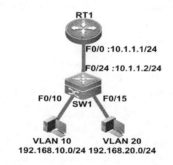

图5-42 基于类的加权公平队列配置

2. 网络设备配置

（1）配置SW1交换机的IP地址

SW1(config)#vlan 10
SW1(config-vlan)#exit
SW1config)#vlan 20
SW1(config-vlan)#exit
SW1(config)#interface range fastEthernet 0/6-10
SW1(config-if-range)#switchport access vlan 10
SW1(config-if-range)#exit
SW1(config)#interface range fastEthernet 0/11-15
SW1(config-if-range)#switchport access vlan 20
SW1(config-if-range)#exit
SW1(config)#interface vlan 10
SW1(config-VLAN 10)#ip address 192.168.10.1 255.255.255.0
SW1(config-VLAN 10)#exit
SW1(config)#interface vlan 20
SW1(config-VLAN 20)#ip address 192.168.20.1 255.255.255.0
SW1(config-VLAN 20)#exit
SW1(config)#interface fastEthernet 0/24
SW1(config-if)#no switchport
SW1(config-if)#ip address 10.1.1.2 255.255.255.0

（2）在SW1上配置静态路由

SW1(config)#ip route 0.0.0.0 0.0.0.0 10.1.1.1

（3）定义协议列表

SW1(config)#access-list 100 permit tcp any any eq 80   #设置网络数据包分类规则
SW1(config)#class-map http   #定义HTTP策略
SW1(config-cmap)#match access-group 100   #关联相关的ACL
SW1(config-cmap)#exit
SW1(config)#access-list 101 permit tcp any any eq 20   #定义FTP策略
SW1(config)#access-list 101 permit tcp any any eq 21   #定义FTP策略
SW1(config)#class-map ftp
SW1(config-cmap)#match access-group 101   #关联相关的ACL
SW1(config-cmap)#exit

（4）定义策略

SW1(config)#policy-map aaa   #创建规则映射表

```
SW1(config-pmap)#class http                         #关联对应类映射表
SW1(config-pmap-c)#bandwidth percent 5              #指定数据流分配带宽
 SW1(config-pmap)#class ftp
SW1(config-pmap-c)#bandwidth percent 25             #指定数据流分配带宽
```
（5）将CBWFQ应用到接口
```
SW1(config)#interface fastEthernet 0/24
SW1(config-if)#service-policy output aaa            #应用服务规则
```
（6）RT1路由器基本配置
```
RT1(config)#interface fastEthernet 0/0
RT1(config-if)#ip address 10.1.1.1 255.255.255.0
RT1(config-if)#no shutdown
RT1(config-if)#exit
RT1(config)#ip route 0.0.0.0 0.0.0.0 10.1.1.2
```
（7）验证测试
```
SW1#show policy-map                                 #查看策略
 Policy Map wy
    Class http
   police cir 10000 bc 1500
      conform-action drop
      exceed-action drop
    Class ftp
     police cir 10000 bc 1500
      conform-action drop
      exceed-action drop

 Policy Map aaa
    Class http
  Bandwidth 5 (%) Max Threshold 64 (packets)
    Class ftp
  Bandwidth 25 (%) Max Threshold 64 (packets)
```

任务4：Rate-limit（流量监管）

任务实施

1. 任务描述及网络拓扑设计

RT1为主校区出口路由器，SW1为主校区汇聚层交换机。在RT1指定F0/0的入接口方向总带宽为50MB。绘制拓扑结构图，如图5-43所示。

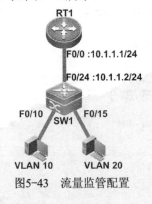

图5-43　流量监管配置

第5章 网络服务应用

2. 网络设备配置

（1）SW1交换机配置

1）在SW1交换机上配置端口隔离、SVI地址。

SW1(config)#vlan 10
SW1(config-vlan)#exit
SW1(config)#vlan 20
SW1(config-vlan)#exit
SW1(config)#interface range fastEthernet 0/6-10
SW1(config-if-range)#switchport access vlan 10
SW1(config-if-range)#exit
SW1(config)#interface range fastEthernet 0/11-15
SW1(config-if-range)#switchport access vlan 20
SW1(config-if-range)#exit
SW1(config)#interface vlan 10
SW1(config-VLAN 10)#ip address 192.168.10.1 255.255.255.0
SW1(config-VLAN 10)#exit
SW1(config)#interface vlan 20
SW1(config-VLAN 20)#ip address 192.168.20.1 255.255.255.0
SW1(config-VLAN 20)#exit
SW1(config)#interface fastEthernet 0/24
SW1(config-FastEthernet 0/24)#no switchport
SW1(config-FastEthernet 0/24)#ip address 10.1.1.2 255.255.255.0
SW1(config-FastEthernet 0/24)#exit

2）在SW1交换机上配置默认路由。

SW1(config)#ip route 0.0.0.0 0.0.0.0 10.1.1.1

（2）RT1路由器基本配置

1）在RT1路由器上配置流量监管。

RT1(config)#interface fastEthernet 0/0
RT1(config-if-FastEthernet 0/0)#ip address 10.1.1.1 255.255.255.0
RT1(config-if-FastEthernet 0/0)#rate-limit input 50000000 14000 14000 conform-action transmit exceed-action drop
　　　　　　　　　　　　　　　　#配置出接口带宽为50MB
RT1(config-if-FastEthernet 0/0)#exit

2）在RT1路由器上配置默认路由。

RT1(config)#ip route 0.0.0.0 0.0.0.0 10.1.1.2

（3）验证测试

RT1#show rate-limit
　FastEthernet 0/0
　　Input
　　matches all traffic
　　params: 50000000 bps, 25000 limit, 25000 extended limit
　　conformed 0 packets, 0 bytes; action: transmit
　　exceeded 0 packets, 0 bytes; action: drop
　　cbucket 50000, cbs 50000; ebucket 0 ebs 0

小结

CQ队列可以按照所示希望的流量定制通过接口的先后顺序让一些重要的流量优先通

过，在网络拥塞的时候可以达到很好的分配流量先后的效果。

PQ队列可以按照所希望的流量定制其通过接口的先后顺序，不仅可以定义相应的网段，还可以定义相应的协议。

CBWFQ带宽控制命令，可以限制通过接口的网段，协议所使用的带宽。

Rate-limit带宽控制命令，只能控制通过接口的总带宽，部分功能与CBWFQ类似。

5.6 组播技术

问题描述

随着宽带多媒体网络的发展，各种宽带网络应用层出不穷。IPTV、视频会议、数据和资料分发、网络音频应用、多媒体远程教育等宽带应用都对现有宽带多媒体网络的承载能力提出了挑战。采用单播技术构建的传统网络已经无法满足新兴宽带网络应用在宽带和网络服务质量方面的要求，随之而来的是网络延时、数据丢失等问题。通过引入IP组播技术，有助于解决以上问题。

问题分析

PIM-DM（Protocol Independent Multicast-Dense Mode）属于密集模式的组播路由协议，通常适合于组播组成员相对比较密集的小型网络。

PIM-SM（Protocol Independent Multicast-Sparse Mode）属于稀疏模式的组播路由协议，通常适合于组播组成员相对分散、范围较广的大中型网络。

配置组播时，主要涉及密集模式独立组播协议（PIM-DM）配置和稀疏模式独立组播协议（PIM-SM）配置。

1. PIM-DM基本配置

步骤1 启动组播路由转发。
router(config)#ip multicast-routing
步骤2 启动PIM-DM。
router(config-if)#ip pim dense-mode
步骤3 配置Hello消息发送间隔。seconds单位为秒。
router(config-if)#**ip pim query-interval** seconds
步骤4 配置PIM邻居过滤。
router(config-if)#**ip pim neighbor-filter** access-list
步骤5 配置PIM状态更新功能。
router(config)#ip pim state-refresh disable
步骤6 配置PIM状态更新消息发送间隔。seconds是1～100的一个整数值。

第5章 网络服务应用

router(config)#ip pim state-refresh origination-interval *seconds*

2. PIM-SM基本配置

步骤1 启动组播路由转发。

router(config)#ip multicast-routing

步骤2 启动PIM-SM。

router(config-if)#ip pim sparse-mode

步骤3 配置静态RP。配置本设备的静态RP。

router(config)#**ip pim rp-address** *rp-address* [*access-list*]

任 务 单

1	PIM-SM（稀疏模式）
2	PIM-DM（密集模式）

解决步骤

根据任务单的安排完成任务。

任务1：PIM-SM（稀疏模式）

任务实施

1. 任务描述及网络拓扑设计

RT1为主校区出口路由器，SW1和SW2为主校区汇聚层交换机。PC1为主校区的视频服务器，为了优化视频流量在主校区的设备上配置组播协议，从而达到优化视频流的目的。绘制拓扑结构图，如图5-44所示。

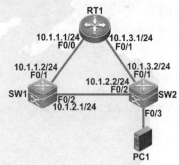

图5-44　PIM-SPARSE模式配置

2. 网络设备配置

（1）SW1交换机配置

1）配置SW1交换机的IP地址。

SW1(config)#interface fastEthernet 0/1
SW1(config-FastEthernet 0/1)#no switchport
SW1(config-FastEthernet 0/1)#ip address 10.1.1.2 255.255.255.0

SW1(config-FastEthernet 0/1)#exit
SW1(config)#interface fastEthernet 0/2
SW1(config-FastEthernet 0/2)#no switchport
SW1(config-FastEthernet 0/2)#ip address 10.1.2.1 255.255.255.0
SW1(config-FastEthernet 0/2)#exit

2）在SW1交换机配置路由。

SW1(config)#router rip
SW1(config-router)#version 2
SW1(config-router)#network 10.1.1.0
SW1(config-router)#network 10.1.2.0
SW1(config-router)#no auto-summary

3）开启组播功能。

SW1(config)#ip multicast-routing   #开启组播路由功能
SW1(config)#interface range fastEthernet 0/1-2
SW1(config-if-range)#ip pim sparse-mode   #启动PIM-SM
SW1(config-if-range)#exit
SW1(config)#ip pim rp-address 10.1.1.1
#在交换机上指定RP，组播源将组播转发到RP，由RP进行路径选择与转发，组播接收者向RP申请组播

（2）SW2交换机配置

1）配置SW2交换机的IP地址。

SW2(config)#vlan 10
SW2(config-vlan)#exit
SW2(config)#interface fastEthernet 0/3
SW2(config-FastEthernet 0/3)#switchport access vlan 10
SW2(config-FastEthernet 0/3)#exit
SW2(config)#interface vlan 10
SW2(config-VLAN 10)#ip address 192.168.100.1 255.255.255.0
SW2(config-VLAN 10)#exit
SW2(config)#interface fastEthernet 0/1
SW2(config-FastEthernet 0/1)#no switchport
SW2(config-FastEthernet 0/1)#ip address 10.1.3.2 255.255.255.0
SW2(config-FastEthernet 0/1)#exit
SW2(config)#interface fastEthernet 0/2
SW2(config-FastEthernet 0/2)#no switchport
SW2(config-FastEthernet 0/2)#ip address 10.1.2.2 255.255.255.0
SW2(config-FastEthernet 0/2)#exit

2）在SW2交换机配置路由。

SW2(config)#router rip
SW2(config-router)#version 2
SW2(config-router)#network 10.1.2.0
SW2(config-router)#network 10.1.3.0
SW2(config-router)#network 192.168.100.0
SW2(config-router)#no auto-summary

3）开启组播功能。

SW2(config)#ip multicast-routing   #开启组播路由功能
SW2(config)#interface range fastEthernet 0/1-2
SW2(config-if-range)#ip pim sparse-mode

```
SW2(config-if-range)#exit
SW2(config)#interface vlan 10
SW2(config-VLAN 10)#ip pim sparse-mode                    #启动PIM-SM
SW2(config-VLAN 10)#exit
SW2(config)#ip pim rp-address 10.1.1.1                    #指定RP
SW2(config)#exit
```

（3）RT1路由器配置。

1）配置RT1路由器IP地址。

```
RT1(config)#interface fastEthernet 0/0
RT1(config-if-FastEthernet 0/0)#ip address 10.1.1.1 255.255.255.0
RT1(config-if-FastEthernet 0/0)#exit
RT1(config)#interface fastEthernet 0/1
RT1(config-if-FastEthernet 0/1)#ip address 10.1.3.1 255.255.255.0
RT1(config-if-FastEthernet 0/1)#exit
```

2）在RT1路由器上配置路由。

```
RT1(config)#router rip
RT1(config-router)#version 2
RT1(config-router)#network 10.1.1.0
RT1(config-router)#network 10.1.3.0
RT1(config-router)#no auto-summary
```

3）开启组播功能。

```
RT1(config)#ip multicast-routing                          #开启组播路由功能
RT1(config)#interface range fastEthernet 0/0-1
RT1(config-if-range)#ip pim sparse-mode                   #启动PIM-SM
RT1(config-if-range)#exit
RT1(config)#ip pim rp-address 10.1.1.1                    #指定RP
RT1(config)#exit
```

（4）验证测试

```
SW1#show ip pim sparse-mode neighbor
Neighbor        Interface              Uptime/Expires    Ver      DR
Address                                                           Priority/Mode
10.1.1.1        FastEthernet 0/1       00:01:49/00:01:37  v2      1 /
10.1.2.2        FastEthernet 0/2       00:01:42/00:01:26  v2      1 / DR
SW1#show ip pim sparse-mode rp mapping
PIM Group-to-RP Mappings
Group(s): 224.0.0.0/4, Static
  RP: 10.1.1.1(Not self) , Static
      Uptime: 00:20:08
```

任务2. PIM-DM（密集模式）

任务实施

1. 任务描述及网络拓扑设计

RT2为主校区出口路由器，SW1和SW1为主校区汇聚层交换机，PC1为主校区的视频服务器。为了优化流量在主校区的设备上配置组播协议。绘制拓扑结构图，如图5-45所示。

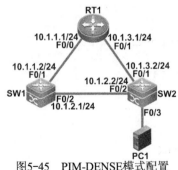

图5-45　PIM-DENSE模式配置

2. 网络设备配置

（1）SW1交换机配置

1）配置SW1交换机的IP地址。

SW1(config)#interface fastEthernet 0/1
SW1(config-FastEthernet 0/1)#no switchport
SW1(config-FastEthernet 0/1)#ip address 10.1.1.2 255.255.255.0
SW1(config-FastEthernet 0/1)#exit
SW1(config)#interface fastEthernet 0/2
SW1(config-FastEthernet 0/2)#no switchport
SW1(config-FastEthernet 0/2)#ip address 10.1.2.1 255.255.255.0
SW1(config-FastEthernet 0/2)#exit

2）在SW1交换机配置路由。

SW1(config)#router rip
SW1(config-router)#version 2
SW1(config-router)#network 10.1.1.0
SW1(config-router)#network 10.1.2.0
SW1(config-router)#no auto-summary

3）开启组播功能。

SW1(config)#ip multicast-routing   #开启组播路由功能
SW1(config)#interface range fastEthernet 0/1-2
SW1(config-if-range)#ip pim dense-mode   #启动PIM-DM
SW1(config-if-range)#exit

（2）SW2交换机配置

1）配置SW2交换机的IP地址。

SW2(config)#vlan 10
SW2(config-vlan)#exit
SW2(config)#interface fastEthernet 0/3
SW2(config-FastEthernet 0/3)#switchport access vlan 10
SW2(config-FastEthernet 0/3)#exit
SW2(config)#interface vlan 10
SW2(config-VLAN 10)#ip address 192.168.100.1 255.255.255.0
SW2(config-VLAN 10)#exit
SW2(config)#interface fastEthernet 0/1
SW2(config-FastEthernet 0/1)#no switchport
SW2(config-FastEthernet 0/1)#ip address 10.1.3.2 255.255.255.0
SW2(config-FastEthernet 0/1)#exit

```
SW2(config)#interface fastEthernet 0/2
SW2(config-FastEthernet 0/2)#no switchport
SW2(config-FastEthernet 0/2)#ip address 10.1.2.2 255.255.255.0
SW2(config-FastEthernet 0/2)#exit
```

2）在SW2交换机配置路由。

```
SW2(config)#router rip
SW2(config-router)#version 2
SW2(config-router)#network 10.1.2.0
SW2(config-router)#network 10.1.3.0
SW2(config-router)#network 192.168.100.0
SW2(config-router)#no auto-summary
```

3）开启组播功能。

```
SW2(config)#ip multicast-routing                              #开启组播路由功能
SW2(config)#interface range fastEthernet 0/1-2
SW2(config-if-range)#ip pim dense-mode                        #启动PIM-DM
SW2(config-if-range)#exit
SW2(config)#interface vlan 10
SW2(config-VLAN 10)#ip pim dense-mode                         #启动PIM-DM
SW2(config-VLAN 10)#exit
SW2(config)#exit
```

（3）RT1路由器配置

1）配置RT1路由器IP地址。

```
RT1(config)#interface fastEthernet 0/0
RT1(config-if-FastEthernet 0/0)#ip address 10.1.1.1 255.255.255.0
RT1(config-if-FastEthernet 0/0)#exit
RT1(config)#interface fastEthernet 0/1
RT1(config-if-FastEthernet 0/1)#ip address 10.1.3.1 255.255.255.0
RT1(config-if-FastEthernet 0/1)#exit
```

2）在RT1路由器上配置路由。

```
RT1(config)#router rip
RT1(config-router)#version 2
RT1(config-router)#network 10.1.1.0
RT1(config-router)#network 10.1.3.0
RT1(config-router)#no auto-summary
```

3）开启组播功能。

```
RT1(config)#ip multicast-routing                              #开启组播路由功能
RT1(config)#interface range fastEthernet 0/0-1
RT1(config-if-range)#ip pim dense-mode                        #启动PIM-DM
RT1(config-if-range)#exit
RT1(config)#exit
```

（4）验证测试

```
    SW1#show ip pim dense-mode neighbor                       #查看PIM邻居
Neighbor-Address    Interface               Uptime/Expires       Ver
10.1.1.1            FastEthernet 0/1        00:06:03/00:01:42    v2
10.1.2.2            FastEthernet 0/2        00:05:50/00:01:31    v2
```

小结

组播是为了更好地分配网络中的大流量数据。由于本实验是为了更好地分配视频流量，所以采用组播协议。

密集模式下的组播流量的传输，如果没有组成员，则自动修剪组播发送信息。稀疏模式下的组播流量传输，只有请求加入组才能获得组播流量。稀疏模式克服了密集模式的低效缺点，不必在开始时将流量泛洪到整个网络，进而再根据需要修剪多播分布树。

5.7 IPv6技术

问题描述

IPv4在过去的几十年中取得了辉煌的成绩，但是由于网络用户及应用快速发展，出现了迫在眉睫的IP地址空间耗尽问题，而这也直接限制了IP技术应用的进一步发展。除了地址短缺外，网络安全性、QoS、性能提高等要求都是IPv4协议本身无法解决的实际问题，因而迫切需要IPv6协议从根本解决目前的问题，同时能够实现网络的无缝过渡。

问题分析

当前Internet上使用IPv4的路由器数量太大，向IPv6过渡只能采用逐步推进的方法，这就要求IPv6系统必须能与IPv4兼容，能够接收和发送IPv4数据包。

IPv4向IPv6过渡的方法主要有两种：一是双协议栈技术，二是隧道技术。

利用双栈策略实现IPv4和IPv6网络的互通，其工作原理是双栈节点同时支持IPv4和IPv6节点的通信，当和IPv4节点通信时需要采用IPv4协议栈，当和IPv6节点通信时需要采用IPv6协议栈。

隧道策略实现IPv4和IPv6网络的互通，其工作原理是将IPv6数据包封装在IPv4数据包中，实现在IPv4网络中的数据传送；同样，IPv4也可以封装在IPv6包中，通过IPv6网络传递到对端的IPv4主机。

配置IPV6主要涉及双协议栈配置和隧道技术配置。用于IPv6穿越 IPv4网络的隧道技术主要有IPv6手工配置隧道、6to4自动隧道、ISATAP自动隧道、IPv6 over IPv4 GRE隧道、6PE隧道。

IPv6基本配置步骤如下：

步骤1　打开接口的IPv6协议。如果没有执行这条命令，给接口配置IPv6地址时，会自动打开IPv6协议。

switch(config-if)#**ipv6 enable**

步骤2　为接口配置IPv6单播地址。ipv6-prefix表示IPv6的网络号，prefix-length 表示IPv6前缀的长度，eui-64表示生成的IPv6地址由配置的地址前缀和64bit的接口ID标识符组成。

switch(config-if)#**ipv6 address** *ipv6-prefix/prefix-length* **[eui-64]**

步骤3　开启IPv6的单播转发功能。

switch(config)#ipv6 unicast-routing

第5章 网络服务应用

任 务 单

1	配置IPv6/IPv4双协议栈
2	配置IPv6 over IPv4 GRE隧道

任务1：配置IPv6/IPv4双协议栈

任务实施

1. 任务描述及网络拓扑设计

RT1为主校区出口路由器，RT2为分校区出口路由器。在两台路由器上启用IPv6/IPv4双协议栈，实现网络的互联互通，使得内网既能使用IPv4的应用，也能使用IPv6的应用。绘制拓扑结构图，如图5-46所示。

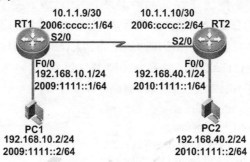

图5-46　IPv6/IPv4双协议栈

2. 网络设备配置

（1）RT1路由器配置

1）开启RT1路由器的IPv6流量转发。

RT1(config)#ipv6 unicast-routing　　　　　　　　　　#开启IPv6流量转发

2）配置RT1路由器的IP地址。

RT1(config)#interface Serial 2/0
RT1(config-if-Serial 2/0)#ip address 10.1.1.9 255.255.255.252　　#设置接口IPv4地址
RT1(config-if-Serial 2/0) #ipv6 enable
RT1(config-if-Serial 2/0)#ipv6 address 2006:CCCC::1/64　　#设置接口IPv6地址
RT1(config)#interface FastEthernet 0/0
RT1(config-if-FastEthernet 0/0)#ip address 192.168.10.1 255.255.255.0
RT1(config-if-FastEthernet 0/0) #ipv6 enable
RT1(config-if-FastEthernet 0/0)#ipv6 address 2009:1111::1/64　　#设置接口IPv6地址
RT1(config-if-FastEthernet 0/0)#exit

3）配置RT1路由器的默认路由。

RT1(config)#ip route 0.0.0.0 0.0.0.0 10.1.1.10　　　　#配置IPv4默认路由
RT1(config)#ipv6 route ::/0 2006:CCCC::2　　　　#配置IPv6默认路由

（2）RT2路由器配置

1）开启RT2路由器的IPv6流量转发。

RT2(config)#ipv6 unicast-routing　　　　　　　　　　#开启IPv6流量转发

2）配置RT2路由器的IP地址。

```
RT2(config)#interface Serial 2/0
RT2(config-if-Serial 2/0)#ip address 10.1.1.10 255.255.255.252      #设置接口IPv4地址
RT2(config-if-Serial 2/0) #ipv6 enable
RT2(config-if-Serial 2/0)# ipv6 address 2006:CCCC::2/64             #设置接口IPv6地址
RT2(config)#interface FastEthernet 0/0
RT2(config-if-FastEthernet 0/0)#ip address 192.168.40.1 255.255.255.0
RT2(config-if-FastEthernet 0/0) #ipv6 enable
RT2(config-if-FastEthernet 0/0)#ipv6 address 2010:1111::1/64
RT2(config-if-FastEthernet 0/0)#exit
```

3）配置RT2路由器的默认路由。

```
RT2(config)#ip route 0.0.0.0 0.0.0.0 10.1.1.9                       #配置IPv4默认路由
RT2(config)#ipv6 route ::/0 2006:CCCC::1                            #配置IPv6默认路由
```

（3）验证测试

按拓扑图，给PC1主机配置IPv4（win7系统，XP系统需要手动安装IPv6包）地址为192.168.10.2/24，网关为192.168.10.1，配置相应的Ipv6地址为2009:1111::2/64，网关为2009:1111::1/64；给PC2主机配置的相应IPv4地址为192.168.40.2/24，网关为192.168.40.1，配置的相应Ipv6地址为2010:1111::2/64，网关为2010:1111::1/64。

1）PC1使用IPV4地址ping PC2，测试效果如下。

```
C:\Documents and Settings\Administrator>ping 192.168.40.2
Pinging 192.168.10.2with 32 bytes of data:
Reply from 192.168.40.2: bytes=32 time<1ms TTL=254
Reply from 192.168.40.2: bytes=32 time<1ms TTL=254
Reply from 192.168.40.2: bytes=32 time<1ms TTL=254
Reply from 192.168.40.2: bytes=32 time<1ms TTL=254
Ping statistics for 192.168.40.2:
    Packets: Sent = 4, Received = 4, Lost = 0(0% loss)
```

2）PC1使用Ipv6地址ping PC2，测试效果如下。

```
C:\Documents and Settings\Administrator>ping 2010:1111::2
正在 Ping 2010:1111::2 具有 32 字节的数据:
来自 2010:1111::2 的回复: 时间<1ms
来自 2010:1111::2 的回复: 时间<1ms
来自 2010:1111::2 的回复: 时间<1ms
来自 2010:1111::2的回复: 时间<1ms

2010:1111::2 的 Ping 统计信息:
    数据包: 已发送=4, 已接收=4, 丢失=0 (0%丢失),
    往返行程的估计时间(以毫秒为单位):
        最短 = 0ms, 最长 = 0ms, 平均 = 0ms
```

任务2: 配置IPv6 over IPv4 GRE隧道

任务实施

1. 任务描述及网络拓扑设计

RT1为主校区出口路由器，RT2为分校区出口路由器。在两台路由器上启用IPv6 over IPv4 GRE隧道策略，实现网络的互联互通。绘制拓扑结构图，如图5-47所示。

第5章 网络服务应用

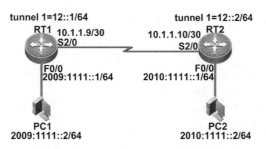

图5-47　IPv6 over IPv4 GRE隧道

2. 网络设备配置

（1）RT1路由器配置

1）开启RT1路由器的IPv6流量转发。

RT1(config)#ipv6 unicast-routing　　　　　　　　　　　#开启IPv6流量转发

2）配置RT1路由器的IP地址。

RT1(config)#interface Serial 2/0
RT1(config-if-Serial 2/0)#ip address 10.1.1.9 255.255.255.252　　#设置接口IPv4地址
RT1(config)#interface FastEthernet 0/0
RT1(config-if-FastEthernet 0/0)# ipv6 address 2009:1111::1/64　　#设置接口Ipv6地址

3）配置IPv6 over IPv4 GRE隧道。

RT1(config)#interface tunnel 1　　　　　　　　　　　#创建隧道接口
RT1(config-if)#ipv6 address 12::1/64　　　　　　　　　#配置隧道接口IPv6地址
RT1(config-if)#tunnel source 10.1.1.9　　　　　　　　　#设置隧道的源地址
RT1(config-if)#tunnel destination 10.1.1.10　　　　　　　#设置隧道的目的地址
RT1(config-if)#tunnel mode gre ipv6　　　　　　　　　#设置隧道封装模式为gre
RT1(config-if)#exit
　　　RT1(config)#ipv6 route ::/0 12::2　　　　　　　　#配置IPv6默认路由

（2）RT2路由器配置

1）开启RT2路由器的IPv6流量转发。

RT2(config)#ipv6 unicast-routing　　　　　　　　　　　#开启IPv6路由功能

2）配置RT2路由器的IP地址。

RT2(config)#interface Serial 2/0
RT2(config-if-Serial 2/0)#ip address 10.1.1.10 255.255.255.252
RT2(config-if-Serial 2/0)#exit
RT2(config)#interface FastEthernet 0/0
RT2(config-if-FastEthernet 0/0)#ipv6 address 2010:1111::1/64

3）配置IPv6 over IPv4 GRE隧道。

RT2(config)#interface tunnel 1　　　　　　　　　　　#创建隧道接口
RT2(config-if)#ipv6 address 12::2/64　　　　　　　　　#配置隧道接口IPv6地址
RT2(config-if)#tunnel source 10.1.1.10　　　　　　　　#设置隧道的源地址
RT2(config-if)#tunnel destination 10.1.1.9　　　　　　　#设置隧道的目的地址
RT2(config-if)#tunnel mode gre ipv6　　　　　　　　　#设置隧道封装模式为gre
RT2(config-if)#exit
RT2(config)#ipv6 route ::/0 12::1

（3）验证测试

按拓扑图，给PC1主机配置的相应IPv6地址为2009:1111::2/64，网关为2009:1111::1/64；

给PC2主机配置的相应IPv6地址为2010:1111::2/64，网关为2010:1111::1/64。PC1使用IPv6地址ping PC2，测试效果如下。

C:\Documents and Settings\Administrator>ping 2010:1111::2
正在 Ping 2010:1111::2 具有 32 B的数据：
来自 2010:1111::2 的回复: 时间<1ms
来自 2010:1111::2 的回复: 时间<1ms
来自 2010:1111::2 的回复: 时间<1ms
来自 2010:1111::2的回复: 时间<1ms

2010:1111::2 的 Ping 统计信息：
　　数据包: 已发送 = 4，已接收 = 4，丢失 = 0 (0% 丢失)，
往返行程的估计时间(以毫秒为单位)：
　　最短 = 0ms，最长 = 0ms，平均 = 0ms

小结

配置IPv6/IPv4双协议栈配置和IPv6 over IPv4 GRE隧道技术时，必须要在路由器上开启IPv6的流量转发。

IPv6/IPv4双协议栈操作和规划复杂，需要同时在网上部署两套路由协议，因为管理上的原因无法大规模部署，所以只能在网络的边缘部署。

利用隧道策略实现IPv4和IPv6网络的互通时，成对隧道部署的要求大大限制了实际的部署范围和部署能力，网络规模不能扩大是一致命缺陷。

第6章 网络工程项目规划与搭建

背景及网络规划

某中等职业学校现有师生员工5000余人,分布在两个不同的校区,为了实现快捷的信息交流和资源共享,需要构建一个跨越两地的校园网络。该学校采用双核心的网络架构和双出口的网络接入模式,该单位主校区对外为100Mbit/s电信链路和1000Mbit/s教育网链路,分校区对外为100Mbit/s电信链路。

主校区主要有教学中心、实训中心、行政办公中心、信息中心(服务器群)四大区域,分校区主要有教学中心、实训中心、行政办公中心、无线区域四大区域。学校为了安全管理每个部门的用户,使用VLAN技术将每个部门的用户划分到不同的VLAN中。分校区采用路由器接入互联网和城域网,分校区的行政办公中心用户采用了无线接入方式,更便于访问网络资源。

为了保障主校区网络的稳定性和拓扑快速收敛,在IP选路中采用的是动态路由协议,因为网络规模较大,所以采用的动态路由协议是开放式最短路径优先(OSPF)。

为了保障主校区与分校区业务数据流传输的高可用性,采用城域网专用链路为主链路,采用基于IPSec VPN技术的互联网链路为备份链路,实现业务流量的高可用性。

要求对局域网进行合理网络规划,要求各大区域能使区域内所有设备均能在通过学校的网关后访问Internet资源,在局域网内部能够实现互联,实现全网的互通。在网络建设中,保证整个网络系统的高性能、高可靠性、高稳定性、高灵活性和高综合性。

校园网络拓扑结构,如图6-1所示。

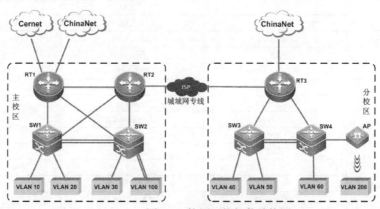

图6-1 校园网络拓扑结构图

具体设备连接关系,见表6-1。

表6-1　网络设备连接表

设备1	设备2	设备1端口	设备2端口	线缆类型
RT1	RT2	S2/0	S2/0	V.35
RT1	SW1	F0/0	F0/1	双绞线
RT1	SW2	F0/1	F0/1	双绞线
RT2	RT3	S3/0	S3/0	V.35
RT2	SW1	F0/0	F0/2	双绞线
RT2	SW1	F0/1	F0/2	双绞线
RT3	SW3	F0/0	F0/1	双绞线
RT3	SW4	F0/1	F0/1	双绞线
SW1	SW1	F0/23-24	F0/23-24	双绞线
SW3	SW4	F0/23-24	F0/23-24	双绞线
SW4	AP	F0/22	G0/1	双绞线

IP地址规划，见表6-2。

表6-2　IP地址规划表

网络设备名称	ID	IP地址	备注
RT1	F0/0	10.1.1.1/30	
	F0/1	10.1.1.5/30	
	S2/0	10.1.1.9/30	
	S3/0	210.21.1.2/24	
	S4/0	68.1.1.2/28	
RT2	F0/0	10.1.1.13/30	
	F0/1	10.1.1.17/30	
	S2/0	10.1.1.10/30	
	S3/0	10.1.1.25/30	
RT3	F0/0	10.1.1.29/30	
	F0/1	10.1.1.33/30	
	S2/0	72.1.1.2/30	
	S3/0	10.1.1.26/30	
SW1	F0/1	10.1.1.2/30	
	F0/2	10.1.1.14/30	
	F0/23-24	10.1.1.21/30	
	VLAN 10	192.168.10.1/24	F0/6-10
	VLAN 20	192.168.20.1/24	F0/11-15
SW2	F0/1	10.1.1.6/30	
	F0/2	10.1.1.18/30	
	F0/23-24	10.1.1.22/30	
	VLAN 30	192.168.30.1/24	F0/6-10
	VLAN 100	192.168.100.1/24	F0/11-15
SW3	F0/1	10.1.1.30/30	
	F0/23-24	10.1.1.41/30	
	VLAN 40	192.168.40.1/24	F0/6-10
	VLAN 50	192.168.50.1/24	F0/11-15
SW4	F0/1	10.1.1.34/30	
	F0/22	10.1.1.37/30	
	F0/23-24	10.1.1.42/30	
	VLAN 60	192.168.60.1/24	F0/6-15
AP	VLAN 200	192.168.200.1/24	无线区域
	G0/1	10.1.1.38/30	

第6章　网络工程项目规划与搭建

项目具体需求

1）主校区和分校区采用VLAN技术，将各个部门的用户主机划分到不同的VLAN中，既可以实现统一管理，又可以保障网络的安全性。主校区教学中心为VLAN 10、实训中心为VLAN 20、行政办公中心为VLAN 30、信息中心为VLAN 100；分校区教学中心为VLAN 40、实训中心为VLAN 50、行政办公中心为VLAN 60、无线办公区域为VLAN 200。服务器群中的服务器主机划分到VLAN 100中。为了便于网络管理，为每个VLAN按照部门名称的汉语拼音为其命名。

2）主校区内部IP选路采用OSPF动态路由协议，区域号为0；主校区与分校区通过城域网进行连接，采用OSPF动态路由协议，区域号为10；为确保OSPF路由协议的稳定性，需要指定RID；为保证OSPF动态路由协议更新安全，需要使用基于接口的MD5验证，验证密码为zhuxiaoqu。

3）分校区内部IP选路采用RIP V2动态路由协议，为保证RIP路由协议更新安全，需要使用基于接口的MD5验证，验证密码是fenxiaoqu。

4）校园网络中存在两个动态路由协议，使用重分发技术实现全网互通，并保障路由更新安全。主校区与分校区连接外网出口存在默认路由，也需要重分发到内网中。

5）电信服务提供商为主校区提供了全局的IP地址段为68.1.1.2/28～68.1.1.14/28，电信服务提供商的IP地址为68.1.1.1/28。使用网络地址转换技术，将私有地址转换为合法的全局IP地址，内部用户可以使用的合法地址为68.1.1.3/28～68.1.1.5/28。教育网服务提供商为主校区提供全局的IP地址段为210.21.1.2/24～210.21.1.254/24，教育网服务提供商的IP地址为210.21.1.1/24。使用网络地址转换技术，将私有地址转换为合法的全局IP地址，内部用户可以使用的合法地址为210.21.1.10/24～210.21.1.15/24。

6）电信服务提供商为分校区提供的全局IP地址为72.1.1.2/30，内部用户用此IP地址段访问互联网，电信服务提供商的IP地址为72.1.1.1/30。

7）将服务器群的FTP服务发布给互联网用户，其合法的全局地址为68.1.1.8/28，内网FTP服务器的地址为192.168.100.4/24；服务器群由两台Web服务器组成Web服务群集组，为了提高服务器的高可用性，要求使用TCP负载均衡来实现，并使用全局地址68.1.1.10/28发布给互联网用户，内网Web服务器的地址为192.168.100.5～192.168.100.6/24。

8）在出口设备上启用策略路由，使内网用户访问教育网资源时走教育网出口，访问其他资源时走电信出口。教育网的资源地址为：211.1.0.0/16、212.1.0.0/24～212.7.0.0/24、213.1.0.0/24～213.1.15.0/24。

9）使用基于时间的访问控制列表，要求该校无线区域用户只能在上班的时间访问互联网（周一至周五的8:00～17:00）。

10）在SW2交换机上设置，该校区所有用户只能在上班的时间才可以访问专用服务器（192.168.100.50/24）的2020端口，实现专门数据库的接口访问。

11）在接入层安全方面，部署端口安全技术，对所有的接入端口配置端口安全，实现交换机接口只允许接入一台主机，如果有违规者，则关闭端口。

12）为了保障主校区与分校区链路安全，需要在主校区与分校区DDN链路上配置PPP，进行先PAP后CHAP的验证方式。将主校区的路由器设置为验证方，客户端的用户名为

fenxiaoqu,密码为123456。

13）为了保障网络链路的带宽，将主校区两台核心交换机相连接的链路配置为链路聚合，并使链路的流量基于源IP地址负载均衡；将分校区两台核心交换机相连接的链路配置为链路聚合，并使链路的流量基于目的IP地址负载均衡。

14）在主校区汇聚层交换机上开启DHCP服务，实现行政办公区域用户主机动态分配IP地址，分配IP地址时，需要为用户主机配置网关和DNS服务器（210.21.1.66）。分校区需要在无线AP上启用DHCP服务，为内部用户主机动态分配IP地址、默认网关和DNS服务器。

15）为保障学校业务的高可用和高可靠性，需要为业务流量配置备份链路，但又要保障数据的安全性，需要部署站点至站点的VPN技术。两校区之间的专用城域网链路为主链路，实现分校区内网与主校区内网通信。如果主链路出现故障，则需要通过互联网使用IPSec VPN技术进行通信，并且要求Gre隧道学习到分校内网的路由。

16）为了方便学校教师远程办公访问内部网络资源，并保障其安全性，需要部署远程接入VPN技术，在路由器的外部接口配置远程接入VPN，允许学校教师采用PPTP访问内网资源，假设只允许用户teacher1、teacher2、teacher3、teacher4和teacher5访问内部网络，其密码为用户名，其获取的IP地址为192.168.30.150~192.168.30.200。

17）为了保障服务器正常工作，防止恶意攻击，需要限制服务器群中的服务器向外发送的最大流量不超过512Kbit/s，并且对超出范围的流量做丢弃处理。

18）给所有三层设备配置Telnet/Enable密码，且只允许192.168.100.240~192.168.100.248网段的管理人员远程登录控制，Telnet密码为123456，Enable密码为654321，密码以明文方式存储。

解决步骤

根据项目具体需求逐步完成任务。首先根据网络拓扑结构图和网络设备连接表，使用以太网线或串口线将网络设备连接起来，对网络设备进行加电，查看设备是否正常工作。然后利用网络设备附带的Console线缆将网络设备的Console端口与主机的串口连接起来，启动网络设备，利用主机上终端软件进行配置管理。

任务1：VLAN配置

1. 任务具体描述

主校区和分校区采用VLAN技术，将各个部门的用户主机划分到不同的VLAN中，既可以实现统一管理，又可以保障网络的安全性。主校区教学中心为VLAN 10、实训中心为VLAN 20、行政办公中心为VLAN 30、信息中心为VLAN 100；分校区教学中心为VLAN 40、实训中心为VLAN 50、行政办公中心为VLAN 60、无线办公区域为VLAN 200。服务器群中的服务器主机划分到VLAN 100中。为了便于网络管理，为每个VLAN按照部门名称的汉语拼音为其命名。网络拓扑结构如图6-1所示。

2. 网络设备配置

（1）在SW1交换机配置VLAN

SW1(config)#vlan 10	#创建VLAN10
SW1(config-vlan)#name jiaoxuezhongxin	#给VLAN10命名
SW1(config-vlan)#exit	
SW1(config)#vlan 20	#创建VLAN20
SW1(config-vlan)#name shixunzhongxin	#给VLAN20命名

（2）在SW2交换机配置VLAN

SW2(config)#vlan 30
SW2(config-vlan)#name xingzhengbangongzhongxin
SW2(config-vlan)#exit
SW2(config)#vlan 100
SW2(config-vlan)#name xinxizhongxin

（3）在SW3交换机配置VLAN

SW3(config)#vlan 40
SW3(config-vlan)#name jiaoxuezhongxin
SW3(config-vlan)#exit
SW3(config)#vlan 50
SW3(config-vlan)#name shixunzhongxin

（4）在SW4交换机配置VLAN

SW4(config)#vlan 60
SW4(config-vlan)#name xingzhengbangongzhongxin
SW4(config-vlan)#exit

（5）在无线AP上配置VLAN

AP(config)#vlan 200	#创建VLAN 200
AP(config-vlan)#name wuxianquyu	#给VLAN200命名

任务2：OSPF路由配置

1. 任务具体描述

主校区内部IP选路采用OSPF动态路由协议，区域号为0；主校区与分校区通过城域网进行连接，采用OSPF动态路由协议，区域号为10；为确保OSPF路由协议的稳定性，需要指定RID；为保证OSPF动态路由协议的更新安全，需要使用基于接口的MD5验证，验证密码为zhuxiaoqu。网络拓扑结构如图6-1所示。

2. 网络设备配置

（1）SW1交换机配置

1）在SW1交换机上配置IP地址。

SW1(config)#interface fastethernet 0/1	#进入接口
SW1(config-if-FastEthernet 0/1)#no switchport	#启用三层功能
SW1(config-if-FastEthernet 0/1)#ip address 10.1.1.2 255.255.255.252	#配置接口IP地址
SW1(config-if-FastEthernet 0/1)#exit	
SW1(config)#interface FastEthernet 0/2	#进入接口
SW1(config-if-FastEthernet 0/2)#no switchport	#启用三层功能
SW1(config-if-FastEthernet 0/2)#ip address 10.1.1.14 255.255.255.252	#配置接口IP地址
SW1(config-if-FastEthernet 0/2)#exit	

```
SW1(config)#interface AggregatePort 1                       #进入接口
SW1(config-if-AggregatePort 1)#no switchport                #启用三层功能
SW1(config-if-AggregatePort 1)#ip address 10.1.1.21 255.255.255.252
                                                            #配置接口IP地址
SW1(config-if-AggregatePort 1)#exit
SW1(config)#interface vlan 10                               #进入虚拟接口
SW1(config-VLAN 10)#ip address 192.168.10.1 255.255.255.0   #配置SVI地址
SW1(config-VLAN 10)#exit
SW1(config)#interface vlan 20                               #进入虚拟接口
SW1(config-VLAN 20)#ip address 192.168.20.1 255.255.255.0   #配置SVI地址
SW1(config-VLAN 20)#exit
```

2）在SW1交换机上配置OSPF路由。

```
SW1(config)#router ospf 1                                   #启用OSPF路由
SW1(config-router)# router-id 1.1.1.1                       #指定路由器ID
SW1(config-router)#network 10.1.1.0 0.0.0.3 area 0          #宣布路由
SW1(config-router)#network 10.1.1.12 0.0.0.3 area 0         #宣布路由
SW1(config-router)#network 10.1.1.20 0.0.0.3 area 0         #宣布路由
SW1(config-router)#network 192.168.10.0 0.0.0.255 area 0    #宣布路由
SW1(config-router)#network 192.168.20.0 0.0.0.255 area 0    #宣布路由
```

3）在SW1交换机上配置OSPF路由协议认证。

```
SW1(config)#interface fastethernet 0/1
SW1(config-if-FastEthernet 0/1)#ip ospf authentication message-digest
                                                            #启用接口的MD5验证
SW1(config-if-FastEthernet 0/1)#ip ospf message-digest-key 1 md5 zhuxiaoqu
                                                            #配置MD5验证的密钥ID和密钥
SW1(config-if-FastEthernet 0/1)#exit
SW1(config)#interface FastEthernet 0/2
SW1(config-if-FastEthernet 0/2)#ip ospf authentication message-digest
                                                            #启用接口的MD5验证
SW1(config-if-FastEthernet 0/2)#ip ospf message-digest-key 1 md5 zhuxiaoqu
                                                            #配置MD5验证的密钥ID和密钥
SW1(config-if-FastEthernet 0/2)#exit
SW1(config)#interface AggregatePort 1
SW1(config-if-AggregatePort 1)#ip ospf authentication message-digest
                                                            #启用接口的MD5验证
SW1(config-if-AggregatePort 1)#ip ospf message-digest-key 1 md5 zhuxiaoqu
                                                            #配置MD5验证的密钥ID和密钥
SW1(config-if-AggregatePort 1)#exit
```

（2）SW2交换机配置

1）在SW2交换机上配置IP地址。

```
SW2(config)#interface FastEthernet 0/1
SW2(config-if-FastEthernet 0/1)#no switchport
SW2(config-if-FastEthernet 0/1)#ip address 10.1.1.6 255.255.255.252
SW2(config-if-FastEthernet 0/1)#exit
SW2(config)#interface FastEthernet 0/2
SW2(config-if-FastEthernet 0/2)#no switchport
SW2(config-if-FastEthernet 0/2)#ip address 10.1.1.18 255.255.255.252
SW2(config-if-FastEthernet 0/2)#exit
SW2(config)#interface AggregatePort 1
```

```
SW2(config-if-AggregatePort 1)#no switchport
SW2(config-if-AggregatePort 1)#ip address 10.1.1.22 255.255.255.252
SW2(config-if-AggregatePort 1)#exit
SW2(config)#interface vlan 30
SW2(config-VLAN 30)#ip address 192.168.30.1 255.255.255.0
SW2(config-VLAN 30)#exit
SW2(config)#interface vlan 100
SW2(config-VLAN 100)#ip address 192.168.100.1 255.255.255.0
SW2(config-VLAN 100)#exit
```

2）在SW2交换机上配置OSPF路由。

```
SW2(config)#router ospf 1
SW2(config-router)# router-id 2.2.2.2
SW2(config-router)#network 10.1.1.4 0.0.0.3 area 0
SW2(config-router)#network 10.1.1.16 0.0.0.3 area 0
SW2(config-router)#network 10.1.1.20 0.0.0.3 area 0
SW2(config-router)#network 192.168.30.0 0.0.0.255 area 0
SW2(config-router)#network 192.168.100.0 0.0.0.255 area 0
```

3）在SW2交换机上配置OSPF路由协议认证。

```
SW2(config)#interface FastEthernet 0/1
SW2(config-if-FastEthernet 0/1)# ip ospf authentication message-digest
SW2(config-if-FastEthernet 0/1)#ip ospf message-digest-key 1 md5 zhuxiaoqu
SW2(config-if-FastEthernet 0/1)#exit
SW2(config)#interface FastEthernet 0/2
SW2(config-if-FastEthernet 0/2)#ip ospf authentication message-digest
SW2(config-if-FastEthernet 0/2)#ip ospf message-digest-key 1 md5 zhuxiaoqu
SW2(config-if-FastEthernet 0/2)#exit
SW2(config)#interface AggregatePort 1
SW2(config-if-AggregatePort 1)#ip ospf authentication message-digest
SW2(config-if-AggregatePort 1)#ip ospf message-digest-key 1 md5 zhuxiaoqu
SW2(config-if-AggregatePort 1)#exit
```

（3）RT1路由器配置

1）在RT1路由器上配置IP地址。

```
RT1(config)#interface Serial 2/0
RT1(config-if-Serial 2/0)#ip address 10.1.1.9 255.255.255.252
RT1(config-if-Serial 2/0)#exit
RT1(config)#interface FastEthernet 0/0
RT1(config-if-FastEthernet 0/0)#ip address 10.1.1.1 255.255.255.252
RT1(config-if-FastEthernet 0/0)#exit
RT1(config)#interface FastEthernet 0/1
RT1(config-if-FastEthernet 0/1)#ip address 10.1.1.5 255.255.255.252
RT1(config-if-FastEthernet 0/1)#exit
```

2）在RT1路由器上配置OSPF路由。

```
RT1(config)#router ospf 1
RT1(config-router)#router-id 3.3.3.3
RT1(config-router)#network 10.1.1.0 0.0.0.3 area 0
RT1(config-router)#network 10.1.1.4 0.0.0.3 area 0
RT1(config-router)#network 10.1.1.8 0.0.0.3 area 0
```

3）在RT1路由器上配置OSPF路由协议认证。

RT1(config)#interface Serial 2/0
RT1(config-if-Serial 2/0)#ip ospf authentication message-digest
RT1(config-if-Serial 2/0)#ip ospf message-digest-key 1 md5 zhuxiaoqu
RT1(config-if-Serial 2/0)#exit
RT1(config)#interface FastEthernet 0/0
RT1(config-if-FastEthernet 0/0)#ip ospf authentication message-digest
RT1(config-if-FastEthernet 0/0)#ip ospf message-digest-key 1 md5 zhuxiaoqu
RT1(config-if-FastEthernet 0/0)#exit
RT1(config)#interface FastEthernet 0/1
RT1(config-if-FastEthernet 0/1)#ip ospf authentication message-digest
RT1(config-if-FastEthernet 0/1)# ip ospf message-digest-key 1 md5 zhuxiaoqu
RT1(config-if-FastEthernet 0/1)#exit

4）在RT1路由器上配置访问外网的默认路由。

RT1(config)#ip route 0.0.0.0 0.0.0.0 68.1.1.1   #配置访问外网的默认路由

（4）RT2路由器配置

1）在RT2路由器上配置IP地址。

RT2(config)#interface Serial 3/0
RT2(config-if-Serial 3/0)#ip address 10.1.1.25 255.255.255.252
RT2(config-if-Serial 3/0)#clock rate 64000
RT2(config-if-Serial 3/0)#exit
RT2(config)#interface Serial 2/0
RT2(config-if-Serial 2/0)#ip address 10.1.1.10 255.255.255.252
RT2(config-if-Serial 2/0)#clock rate 64000
RT2(config-if-Serial 2/0)#exit
RT2(config)#interface FastEthernet 0/0
RT2(config-if-FastEthernet 0/0)#ip address 10.1.1.13 255.255.255.252
RT2(config-if-FastEthernet 0/0)#exit
RT2(config)#interface FastEthernet 0/1
RT2(config-if-FastEthernet 0/1)#ip address 10.1.1.17 255.255.255.252

2）在RT2路由器上配置OSPF路由。

RT2(config)#router ospf 1
RT2(config-router)# router-id 4.4.4.4
RT2(config-router)#network 10.1.1.8 0.0.0.3 area 0
RT2(config-router)#network 10.1.1.12 0.0.0.3 area 0
RT2(config-router)#network 10.1.1.16 0.0.0.3 area 0
RT2(config-router)#network 10.1.1.24 0.0.0.3 area 10

3）在RT2路由器上配置OSPF路由协议认证。

RT2(config)#interface Serial 3/0
RT2(config-if-Serial 3/0)#ip ospf authentication message-digest
RT2(config-if-Serial 3/0)#ip ospf message-digest-key 1 md5 zhuxiaoqu
RT2(config-if-Serial 3/0)#exit
RT2(config)#interface Serial 2/0
RT2(config-if-Serial 2/0)#ip ospf authentication message-digest
RT2(config-if-Serial 2/0)#ip ospf message-digest-key 1 md5 zhuxiaoqu
RT2(config-if-Serial 2/0)#exit
RT2(config)#interface FastEthernet 0/0
RT2(config-if-FastEthernet 0/0)#ip ospf authentication message-digest
RT2(config-if-FastEthernet 0/0)#ip ospf message-digest-key 1 md5 zhuxiaoqu
RT2(config-if-FastEthernet 0/0)#exit

第6章 网络工程项目规划与搭建

```
RT2(config)#interface FastEthernet 0/1
RT2(config-if-FastEthernet 0/1#ip ospf authentication message-digest
RT2(config-if-FastEthernet 0/1)#ip ospf message-digest-key 1 md5 zhuxiaoqu
RT2(config-if-FastEthernet 0/1)#exit
```

（5）RT3路由器配置

1）在RT3路由器上配置IP地址。

```
RT3(config)#interface Serial 3/0
RT3(config-if-Serial 3/0)#ip address 10.1.1.26 255.255.255.252
RT3(config-if-Serial 3/0)#exit
```

2）在RT3路由器上配置OSPF路由。

```
RT3(config)#router ospf 1
RT3(config-router)#router-id 5.5.5.5
RT3(config-router)#network 10.1.1.24 0.0.0.3 area 10
```

3）在RT3路由器上配置OSPF路由协议认证。

```
RT3(config)#interface Serial 3/0
RT3(config-if-Serial 3/0)#ip ospf authentication message-digest
RT3(config-if-Serial 3/0)#ip ospf message-digest-key 1 md5 zhuxiaoqu
RT3(config-if-Serial 3/0)#exit
```

4）在RT3路由器上配置访问外网的默认路由。

```
RT3(config)#ip route 0.0.0.0 0.0.0.0 72.1.1.1          #配置访问外网的默认路由
```

任务3：RIP路由配置

1. 任务具体描述

分校区内部IP选路采用RIP V2动态路由协议，为保证RIP路由协议的更新安全，需要使用基于接口的MD5验证，验证密码是fenxiaoqu。网络拓扑结构如图6-1所示。

2. 网络设备配置

（1）SW3交换机配置

1）在SW3交换机上配置IP地址。

```
SW3(config)#interface FastEthernet 0/1
SW3(config-if-FastEthernet 0/1)#no switchport
SW3(config-if-FastEthernet 0/1)#ip address 10.1.1.30 255.255.255.252
SW3(config-if-FastEthernet 0/1)#exit
SW3(config)#interface AggregatePort 2
SW3(config-if-AggregatePort 2)#no switchport
SW3(config-if-AggregatePort 2)#ip address 10.1.1.41 255.255.255.252
SW3(config-if-AggregatePort 2)#exit
SW3(config)#interface vlan 40
SW3(config-VLAN 40)#ip address 192.168.40.1 255.255.255.0
SW3(config-VLAN 40)#exit
SW3(config)#interface vlan 50
SW3(config-VLAN 50)#ip address 192.168.50.1 255.255.255.0
SW3(config-VLAN 50)#exit
```

2）在SW3交换机上配置RIP路由。

```
SW3(config)#router rip                                 #启用RIP路由进程
SW3(config-router)#version 2                           #定义RIP版本号
```

SW3(config-router)#network 10.0.0.0   #宣布路由
SW3(config-router)#network 192.168.40.0   #宣布路由
SW3(config-router)#network 192.168.50.0   #宣布路由
SW3(config-router)#no auto-summary   #关闭自动汇总

3）在SW3交换机上配置RIP路由协议认证。

SW3(config)#key chain rip   #配置密钥链
SW3(config-keychain)#key 1   #配置密钥ID
SW3(config-keychain-key)#key-string fenxiaoqu   #配置密钥值
SW3(config)#interface FastEthernet 0/1
SW3(config-if-FastEthernet 0/1)#ip rip authentication mode md5   #配置验证方式为MD5
SW3(config-if-FastEthernet 0/1)#ip rip authentication key-chain rip #接口上应用密钥链
SW3(config-if-FastEthernet 0/1)#exit
SW3(config)#interface AggregatePort 2
SW3(config-if-AggregatePort 2)#ip rip authentication mode md5   #配置验证方式为MD5
SW3(config-if-AggregatePort 2)#ip rip authentication key-chain rip #接口上应用密钥链
SW3(config-if-AggregatePort 2)#exit

（2）SW4交换机配置

1）在SW4交换机上配置IP地址。

SW4(config)#interface FastEthernet 0/1
SW4(config-if-FastEthernet 0/1)#no switchport
SW4(config-if-FastEthernet 0/1)#ip address 10.1.1.34 255.255.255.252
SW4(config-if-FastEthernet 0/1)#exit
SW4(config)#interface AggregatePort 2
SW4(config-if-AggregatePort 2)#no switchport
SW4(config-if-AggregatePort 2)#ip address 10.1.1.42 255.255.255.252
SW4(config-if-AggregatePort 2)#exit
SW4(config)#interface vlan 60
SW4(config-VLAN 60)#ip address 192.168.60.1 255.255.255.0
SW4(config-VLAN 60)#exit

2）在SW4交换机上配置RIP路由。

SW4(config)#router rip
SW4(config-router)#version 2
SW4(config-router)#network 10.0.0.0
SW4(config-router)#network 192.168.60.0
SW4(config-router)#no auto-summary

3）在SW4交换机上配置RIP路由协议认证。

SW4(config)#key chain rip
SW4(config-keychain)#key 1
SW4(config-keychain-key)#key-string fenxiaoqu
SW4(config)#interface FastEthernet 0/1
SW4(config-if-FastEthernet 0/1)#ip rip authentication mode md5
SW4(config-if-FastEthernet 0/1)#ip rip authentication key-chain rip
SW4(config-if-FastEthernet 0/1)#exit
SW4(config)#interface AggregatePort 2
SW4(config-if-AggregatePort 2)#ip rip authentication mode md5
SW4(config-if-AggregatePort 2)#ip rip authentication key-chain rip
SW4(config-if-AggregatePort 2)#exit

第6章 网络工程项目规划与搭建

（3）RT3路由器配置

1）在RT3路由器上配置IP地址。

RT3(config)# interface FastEthernet 0/0
RT3(config-if-FastEthernet 0/0)#ip address 10.1.1.29 255.255.255.252
RT3(config-if-FastEthernet 0/0)#exit
RT3(config)#interface FastEthernet 0/1
RT3(config-if-FastEthernet 0/1)#ip address 10.1.1.33 255.255.255.252
RT3(config-if-FastEthernet 0/1)#exit

2）在RT3路由器上配置RIP路由。

RT3(config)#router rip
RT3(config-router)#version 2
RT3(config-router)#network 10.0.0.0
RT3(config-router)#no auto-summary

3）在RT3路由器上配置RIP路由协议认证。

RT3(config)#key chain rip
RT3(config-keychain)#key 1
RT3(config-keychain-key)#key-string fenxiaoqu
RT3(config)# interface FastEthernet 0/0
RT3(config-if-FastEthernet 0/0)#ip rip authentication mode md5
RT3(config-if-FastEthernet 0/0)#ip rip authentication key-chain rip
RT3(config-if-FastEthernet 0/0)#exit
RT3(config)#interface FastEthernet 0/1
RT3(config-if-FastEthernet 0/1)#ip rip authentication mode md5
RT3(config-if-FastEthernet 0/1)#ip rip authentication key-chain rip
RT3(config-if-FastEthernet 0/1)#exit

任务4：路由重分发配置

1. 任务具体描述

校园网络中存在两个动态路由协议，使用重分发技术实现全网互通，并保障路由的更新安全。主校区与分校区连接外网出口存在默认路由，也需要重分发到内网中。网络拓扑结构如图6-1所示。

2. 网络设备配置

（1）在RT1路由器上配置默认路由重分发

RT1(config)#router ospf 1
RT1(config-router)#default-information originate   #重分发默认路由

（2）在RT3路由器上配置RIP路由和默认路由重分发

RT3(config)#router rip
RT3(config-router)#version 2
RT3(config-router)#network 10.0.0.0
RT3(config-router)#no auto-summary
RT3(config-router)#redistribute ospf 1 metric 2   #将OSPF路由重分发到RIP中
RT3(config-router)#default-information originate   #重分发默认路由

（3）在RT3路由器上配置OSPF路由重分发

RT3(config)#router ospf 1
RT3(config-router)#router-id 5.5.5.5   #指定路由器ID

```
RT3(config-router)#redistribute rip subnets              #将RIP路由重分发到OSPF中
RT3(config-router)#network 10.1.1.24 0.0.0.3 area 10    #宣布路由
RT3(config-router)#network 10.10.10.0 0.0.0.255 area 10 #宣布路由
```

任务5：主校区网络地址转换技术配置

1. 任务具体描述

电信服务提供商为主校区提供了全局的IP地址段为68.1.1.2/28～68.1.1.14/28，电信服务提供商的IP地址为68.1.1.1/28。使用网络地址转换技术，将私有地址转换为合法的全局IP地址，内部用户可以使用的合法地址为68.1.1.3/28～68.1.1.5/28。教育网服务提供商为主校区提供全局的IP地址段为210.21.1.2/24～210.21.1.254/24，教育网服务提供商的IP地址为210.21.1.1/24。使用网络地址转换技术，将私有地址转换为合法的全局IP地址，内部用户可以使用的合法地址为210.21.1.10/24～210.21.1.15/24。网络拓扑结构如图6-1所示。

2. 网络设备配置

（1）在RT1路由器配置IP地址和指定内外部接口

```
RT1(config)#interface Serial 2/0
RT1(config-if-Serial 2/0)#ip nat inside                 #定义接口为内部接口
RT1(config-if-Serial 2/0)#exit
RT1(config)#interface FastEthernet 0/0
RT1(config-if-FastEthernet 0/0)#ip nat inside           #定义接口为内部接口
RT1(config-if-FastEthernet 0/0)#exit
RT1(config)#interface FastEthernet 0/1
RT1(config-if-FastEthernet 0/1)#ip nat inside           #定义接口为内部接口
RT1(config-if-FastEthernet 0/1)#exit
RT1(config)#interface Serial 3/0
RT1(config-if-Serial 3/0)#ip nat outside                #定义接口为外部接口
RT1(config-if-Serial 3/0)#ip address 210.21.1.2 255.255.255.0
RT1(config-if-Serial 3/0)#clock rate 64000
RT1(config-if-Serial 3/0)#exit
RT1(config)#interface Serial 4/0
RT1(config-if-Serial 4/0)#ip nat outside                #定义接口为外部接口
RT1(config-if-Serial 4/0)#ip address 68.1.1.2 255.255.255.248
RT1(config-if-Serial 4/0)#clock rate 64000
```

（2）在RT1路由器上配置路由选择

```
RT1(config)#ip route 0.0.0.0 0.0.0.0 68.1.1.1           #配置访问外网默认路由
```

（3）在RT1路由器配置访问控制列表

```
RT1(config)#ip access-list extended 100                 #配置允许访问外网的列表
RT1(config-ext-nacl)#permit ip any any
RT1(config)#access-list 101 permit ip any any           #配置允许访问外网的列表
```

（4）在RT1路由器配置地址池

```
RT1(config)#ip nat pool dianxin 68.1.1.3 68.1.1.5 netmask 255.255.255.248
#配置电信网出口NAT地址池
RT1(config)#ip nat pool jiaoyu 210.21.1.10 210.21.1.15 netmask 255.255.255.0
#配置教育网出口NAT地址池
```

（5）在RT1路由器配置地址转换，将符合条件的内部地址转换到地址池中的全局地址

```
RT1(config)#ip nat inside source list 100 pool dianxin overload
```

第6章 网络工程项目规划与搭建

```
                                                      #配置电信网出口动态NAT，允许内网访问互联网
RT1(config)#ip nat inside source list 101 pool jiaoyu overload
                                                      #配置教育网出口动态NAT，允许内网访问互联网
```

▶ 任务6：分校区网络地址转换技术配置

1. 任务具体描述

电信服务提供商为分校区提供的全局IP地址为72.1.1.2/30，内部用户用此IP地址段访问互联网，电信服务提供商的IP地址为72.1.1.1/30。网络拓扑结构如图6-1所示。

2. 网络设备配置

（1）在RT3路由器配置IP地址和指定内外部接口

```
RT3(config)#interface serial 3/0
RT3(config-if-Serial 3/0)#ip nat inside              #定义接口为内部接口
RT3(config-if-Serial 3/0)#exit
RT3(config)#interface FastEthernet 0/1
RT3(config-if-FastEthernet 0/1)#ip nat inside        #定义接口为内部接口
RT3(config-if-FastEthernet 0/1)#exit
RT3(config)#interface Serial 2/0
RT3(config-if-Serial 2/0)#ip nat outside             #定义接口为外部接口
RT3(config-if-Serial 2/0)#ip address 72.1.1.2 255.255.255.252
```

（2）在RT3路由器上配置路由选择

```
RT3(config)#ip route 0.0.0.0 0.0.0.0 72.1.1.1        #配置访问外网默认路由
```

（3）在RT3路由器配置访问控制列表

```
RT3(config)#ip access-list extended 101              #创建时间访问列表
RT3(config-ext-nacl)#permit ip 192.168.200.0 0.0.0.255 any time-range 1
RT3(config-ext-nacl)# deny ip 192.168.200.0 0.0.0.255 any
RT3(config-ext-nacl)# permit ip any any
```

（4）在RT3路由器配置基于Serial 2/0接口地址转换

```
RT3(config)#ip nat inside source list 101 interface Serial 2/0   #配置基于接口NAT转换
```

▶ 任务7：TCP负载均衡配置

1. 任务具体描述

将服务器群的FTP服务发布给互联网用户，其合法的全局地址为68.1.1.8/28，内网FTP服务器的地址为192.168.100.4/24；服务器群由两台Web服务器组成Web服务群集组，为了提高服务器的高可用性，要求使用TCP负载均衡来实现，并使用全局地址68.1.1.10/28发布给互联网用户，内网Web服务器的地址为192.168.100.5～192.168.100.6/24。网络拓扑结构如图6-1所示。

2. 网络设备配置

（1）在RT1路由器配置FTP服务发布

```
RT1(config)#ip nat inside source static tcp 192.168.100.4 21 68.1.1.8 21
RT1(config)#ip nat inside source static tcp 192.168.100.4 20 68.1.1.8 20   #将FTP服务发布到互联网
```

（2）在RT1路由器配置WWW服务发布

```
RT1(config)#ip access-list extended 100              #定义允许NAT转换的列表
RT1(config-ext-nacl)#permit tcp any host 68.1.1.10
```

RT1(config)#ip nat pool web 192.168.100.5 192.168.100.6 prefix-length 24 type rotary
　　　　　　　　　　　　　　　　　　　　　　　　　　　　#定义TCP负载均衡池
RT1(config)#ip nat inside destination list 100 pool web　　#实现TCP负载分担

任务8：策略路由配置

1. 任务具体描述

在主校区出口设备上启用策略路由，使VLAN 10和VLAN 20用户访问外网走教育网链路；其他用户访问教育网资源时走教育网出口，访问其他资源时走电信出口。教育网的资源地址为：211.1.0.0/16、212.1.0.0/24～212.7.0.0/24、213.1.0.0/24～213.1.15.0/24。网络拓扑结构如图6-1所示。

2. 网络设备配置

（1）在RT1路由器配置策略路由

RT1(config)#ip access-list standard 3　　　　　　　　#定义列表
RT1(config-ext-nacl)#permit 192.168.10.0 0.0.0.255
RT1(config-ext-nacl)#permit 192.168.20.0 0.0.0.255
RT1(config-ext-nacl)#exit
RT1(config)#route-map jiaoyu permit 10　　　　　　　　#配置名为jiaoyu的route-map
RT1(config-route-map)#match ip address 3　　　　　　　#匹配列表的数据执行下面动作
RT1(config-route-map)#set ip next-hop 210.21.1.1　　　#设置下一跳地址为210.21.1.1
RT1(config-route-map)#exit
RT1(config)#interface FastEthernet 0/0
RT1(config-if-FastEthernet 0/0)#ip policy route-map jiaoyu　#应用route-map
RT1(config)#interface FastEthernet 0/1
RT1(config-if-FastEthernet 0/1)#ip policy route-map jiaoyu

（2）在RT1路由器配置访问教育网资源时的静态路由

RT1(config)#ip route 211.1.0.0 255.255.0.0 210.21.1.1　　　　#配置访问教育网静态路由
RT1(config)#ip route 212.1.0.0 255.255.255.0 210.21.1.1　　　#配置访问教育网静态路由
RT1(config)#ip route 212.2.0.0 255.255.255.0 210.21.1.1　　　#配置访问教育网静态路由
RT1(config)#ip route 212.3.0.0 255.255.255.0 210.21.1.1　　　#配置访问教育网静态路由
RT1(config)#ip route 212.4.0.0 255.255.255.0 210.21.1.1　　　#配置访问教育网静态路由
RT1(config)#ip route 212.5.0.0 255.255.255.0 210.21.1.1　　　#配置访问教育网静态路由
RT1(config)#ip route 212.6.0.0 255.255.255.0 210.21.1.1　　　#配置访问教育网静态路由
RT1(config)#ip route 212.7.0.0 255.255.255.0 210.21.1.1　　　#配置访问教育网静态路由
RT1(config)#ip route 213.1.0.0 255.255.255.0 210.21.1.1　　　#配置访问教育网静态路由
RT1(config)#ip route 213.1.1.0 255.255.255.0 210.21.1.1　　　#配置访问教育网静态路由
RT1(config)#ip route 213.1.2.0 255.255.255.0 210.21.1.1　　　#配置访问教育网静态路由
RT1(config)#ip route 213.1.3.0 255.255.255.0 210.21.1.1　　　#配置访问教育网静态路由
RT1(config)#ip route 213.1.4.0 255.255.255.0 210.21.1.1　　　#配置访问教育网静态路由
RT1(config)#ip route 213.1.5.0 255.255.255.0 210.21.1.1　　　#配置访问教育网静态路由
RT1(config)#ip route 213.1.6.0 255.255.255.0 210.21.1.1　　　#配置访问教育网静态路由
RT1(config)#ip route 213.1.7.0 255.255.255.0 210.21.1.1　　　#配置访问教育网静态路由
RT1(config)#ip route 213.1.8.0 255.255.255.0 210.21.1.1　　　#配置访问教育网静态路由
RT1(config)#ip route 213.1.9.0 255.255.255.0 210.21.1.1　　　#配置访问教育网静态路由
RT1(config)#ip route 213.1.10.0 255.255.255.0 210.21.1.1　　　#配置访问教育网静态路由
RT1(config)#ip route 213.1.11.0 255.255.255.0 210.21.1.1　　　#配置访问教育网静态路由
RT1(config)#ip route 213.1.12.0 255.255.255.0 210.21.1.1　　　#配置访问教育网静态路由
RT1(config)#ip route 213.1.13.0 255.255.255.0 210.21.1.1　　　#配置访问教育网静态路由

第6章　网络工程项目规划与搭建

```
RT1(config)#ip route 213.1.14.0 255.255.255.0 210.21.1.1      #配置访问教育网静态路由
RT1(config)#ip route 213.1.15.0 255.255.255.0 210.21.1.1      #配置访问教育网静态路由
```
（3）在RT1路由器配置访问其他资源时的默认路由
```
RT1(config)#ip route 0.0.0.0 0.0.0.0 68.1.1.1
```

任务9：基于时间访问控制列表配置

1. 任务具体描述

使用基于时间的访问控制列表，要求学校无线区域用户只能在上班的时间访问互联网（周一至周五的8:00～17:00）。网络拓扑结构如图6-1所示。

2. 网络设备配置

（1）在RT3路由器配置基于时间的访问控制列表
```
RT3(config)#time-range 1                                       #创建时间段
RT3(config-time-range)#periodic Weekdays 8:00 to 17:00         #定义周期时间
RT3(config-time-range)#exit
RT3(config)#ip access-list extended 101
RT3(config-ext-nacl)#permit ip 192.168.200.0 0.0.0.255 any time-range 1
RT3(config-ext-nacl)#deny ip 192.168.200.0 0.0.0.255 any
RT3(config-ext-nacl)#permit ip any any
```
#上面4条命令是创建访问列表，并应用时间限制。表示在应用时间内，允许192.168.200.0网段访问外网，其余时间不可以访问外网

（2）在RT3路由器配置基于外部端口的NAT转换
```
RT3(config)#ip nat inside source list 101 interface Serial 2/0
```

任务10：访问控制列表配置

1. 任务具体描述

在SW2交换机上设置，校区所有用户只能在上班的时间才可以访问专用服务器（192.168.100.50）的2020端口，实现专门数据库的接口访问。网络拓扑结构如图6-1所示。

2. 网络设备配置

（1）在SW2交换机配置基于时间的访问控制列表
```
SW2(config)#time-range 1                                       #创建时间段
SW2(config-time-range)#periodic Weekdays 8:00 to 17:00         #定义周期时间
SW2(config-time-range)#exit
SW2(config)#ip access-list extended 100
SW2(config-ext-nacl)# permit tcp any host 192.168.100.50 eq 2020 time-range 1
SW2(config-ext-nacl)# permit udp any host 192.168.100.50 eq 2020 time-range 1
SW2(config-ext-nacl)#deny tcp any host 192.168.100.50 eq 2020
SW2(config-ext-nacl)#deny udp any host 192.168.100.50 eq 2020
SW2(config-ext-nacl)# permit ip any any
```
#上面5条命令是创建访问列表，并应用时间限制。表示在应用时间内，校区用户可以访问192.168.100.50服务器的2020端口，实现专门数据库的访问，其余时间不可以访问192.168.100.50服务器的2020端口，最后允许别的所有数据包通过

（2）在VLAN 100的SVI接口上应用ACL
```
SW2(config)#interface VLAN 100
SW2(config-vlan)#ip access-group 100 out
```

任务11：端口安全配置

1. 任务具体描述

在接入层安全方面，部署端口安全技术，对所有的接入端口配置端口安全，实现交换机接口只允许接入一台主机，如果有违规者，则关闭端口。网络拓扑结构如图6-1所示。

2. 网络设备配置

（1）在SW1交换机配置端口安全

SW1(config)#int range FastEthernet 0/6-10   #进入接口范围模式
SW1(config-if-range)#switchport access vlan 10   #端口划分到VLAN
SW1(config-if-range)#switchport port-security   #开启端口安全
SW1(config-if-range)#switchport port-security maximum 1   #配置接口接入数量
SW1(config-if-range)#switchport port-security violation shutdown   #配置违规处理方式
SW1(config-if-range)#exit
SW1(config)#int range FastEthernet 0/11-15   #进入接口范围模式
SW1(config-if-range)#switchport access vlan 20   #端口划分到VLAN
SW1(config-if-range)#switchport port-security   #开启端口安全
SW1(config-if-range)#switchport port-security maximum 1   #配置接口接入数量
SW1(config-if-range)#switchport port-security violation shutdown   #配置违规处理方式

（2）在SW2交换机配置端口安全

SW2(config)#int range FastEthernet 0/6-10
SW2(config-if-range)#switchport access vlan 30
SW2(config-if-range)#switchport port-security
SW2(config-if-range)#switchport port-security maximum 1
SW2(config-if-range)#switchport port-security violation shutdown
SW2(config-if-range)#exit
SW2(config)#int range FastEthernet 0/11-15
SW2(config-if-range)#switchport access vlan 100
SW2(config-if-range)#switchport port-security maximum 1
SW2(config-if-range)#switchport port-security violation shutdown
SW2(config-if-range)#switchport port-security

（3）在SW3交换机配置端口安全

SW3(config)#int range FastEthernet 0/6-10
SW3(config-if-range)#switchport access vlan 40
SW3(config-if-range)#switchport port-security
SW3(config-if-range)#switchport port-security maximum 1
SW3(config-if-range)#switchport port-security violation shutdown
SW3(config-if-range)#exit
SW3(config)#int range FastEthernet 0/11-15
SW3(config-if-range)#switchport access vlan 50
SW3(config-if-range)#switchport port-security maximum 1
SW3(config-if-range)#switchport port-security violation shutdown
SW3(config-if-range)#switchport port-security

（4）在SW4交换机配置端口安全

SW4(config)#int range FastEthernet 0/6-15
SW4(config-if-range)#switchport port-security
SW4(config-if-range)#switchport access vlan 60
SW4(config-if-range)#switchport port-security maximum 1
SW4(config-if-range)#switchport port-security violation shutdown

第6章　网络工程项目规划与搭建

任务12：PPP协议认证配置

1. 任务具体描述

为了保障主校区与分校区链路安全，需要在主校区与分校区DDN链路上配置PPP，进行先PAP后CHAP的验证方式。将主校区的路由器设置为验证方，客户端的用户名为fenxiaoqu，密码为123456。网络拓扑结构如图6-1所示。

2. 网络设备配置

（1）在RT2路由器上配置PPP协议

```
RT2(config)#username fenxiaoqu password 123456                    #创建验证数据库
RT2(config-if-Serial 3/0)#interface Serial 3/0
RT2(config-if-Serial 3/0)#encapsulation PPP                       #封装PPP协议
RT2(config-if-Serial 3/0)#ppp authentication pap chap             #启用PAP CHAP认证
RT2(config-if-Serial 3/0)#ip address 10.1.1.25 255.255.255.252
RT2(config-if-Serial 3/0)#clock rate 64000
```

（2）在RT3路由器上配置PPP协议

```
RT3(config)#interface Serial 3/0
RT3(config-if-Serial 3/0)#encapsulation PPP                       #封装PPP协议
RT3(config-if-Serial 3/0)#ppp chap hostname fenxiaoqu             #指定CHAP验证的主机名
RT3(config-if-Serial 3/0)#ppp chap password 123456                #指定CHAP验证的密码
RT3(config-if-Serial 3/0)#ppp pap sent-username fenxiaoqu password 123456
                     # PAP验证时，客户端将用户名和密码发送到服务端验证
RT3(config-if-Serial 3/0)#ip address 10.1.1.26 255.255.255.252
```

任务13：链路聚合配置

1. 任务具体描述

为了保障网络链路的带宽，将主校区两台核心交换机相连接的链路配置为链路聚合，并使链路的流量基于源IP地址负载均衡；将分校区两台核心交换机相连接的链路配置为链路聚合，并使链路的流量基于目的IP地址负载均衡。网络拓扑结构如图6-1所示。

2. 网络设备配置

（1）在SW1和SW2之间链路配置链路聚合，并实现基于源地址的流量负载平衡

1）SW1交换机配置。

```
SW1(config)#aggregateport load-balance src-ip                     #配置基于源地址负载均衡
```

2）SW2交换机配置。

```
SW2(config)#aggregateport load-balance src-ip                     #配置基于源地址负载均衡
```

（2）在SW3和SW4之间链路配置链路聚合，并实现基于目的IP地址流量负载平衡

1）SW3交换机配置。

```
SW3(config)#aggregateport load-balance dst-ip                     #配置基于目的地址负载均衡
```

2）SW4交换机配置。

```
SW4(config)#aggregateport load-balance dst-ip                     #配置基于目的地址负载均衡
```

任务14：DHCP服务配置

1. 任务具体描述

在主校区汇聚层交换机上开启DHCP服务，实现行政办公区域用户主机动态分配IP地址，分配IP地址时，需要为用户主机配置网关和DNS服务器（210.21.1.66）。分校区需要在无线AP上启用DHCP服务，为内部用户主机动态分配IP地址、默认网关和DNS服务器。网络拓扑结构如图6-1所示。

2. 网络设备配置

（1）在SW2交换机上配置DHCP服务，给行政办公用户动态分配IP地址

SW2(config)#server dhcp	#启用DHCP服务
SW2(config)#ip dhcp pool vlan30	#创建DHCP地址池
SW2(dhcp-config)#network 192.168.30.0 255.255.255.0	#定义地址段
SW2(dhcp-config)#dns-server 210.21.1.66	#定义DNS地址
SW2(dhcp-config)#default-router 192.168.30.1	#定义默认网关
SW2(dhcp-config)#exit	
SW2(config)#ip dhcp excluded-address 192.168.30.1	#定义排除地址

（2）在无线AP上配置DHCP服务

AP(config)#service dhcp	#启用DHCP服务
AP(config)#ip dhcp pool pool1	#创建DHCP地址池
AP(dhcp-config)#network 192.168.200.0 255.255.255.0	#定义地址段
AP(dhcp-config)#dns-server 210.21.1.66	#定义DNS地址
AP(dhcp-config)#default-router 192.168.200.1	#定义默认网关
AP(config)#vlan 200	#创建VLAN 200
AP(config-vlan)#name wuxianquyu	#为VLAN命名
AP(config-vlan)#exit	
AP(config)#dot11 wlan 10	#创建WLAN
AP(config-wlan-config)#vlan 200	#绑定VLAN
AP(config-wlan-config)#no broadcast-ssid	#不广播SSID
AP(config-wlan-config)#ssid wuxianquyu	#定义SSID
AP(config-wlan-config)#exit	
AP(config)#interface Dot11radio 1/0	#进入接口模式
AP(config-if-dot11radio 1/0)#encapstlation dot1Q 200	#封装802.1Q
AP(config-if-dot11radio 1/0)#mac-mode fat	#定义胖AP模式
AP(config-if-dot11radio 1/0)#radio-type 802.11b	#定义radio类型
AP(config-if-dot11radio 1/0)#channel 1	#定义信道
AP(config-if-dot11radio 1/0)#wlan-id 10	#绑定WLAN
AP(config)#interface BVI 200	#进入BVI接口
AP(config-if-BVI 1)#ip address 192.168.200.1 255.255.255.0	#配置接口IP地址
AP(config)#wlansec 10	#进入无线安全配置模式
AP(config)#security rsn enable	#配置开启RSN安全模式
AP(config)#security rsn ciphers aes enable	#配置WAP2的加密模式为AES
AP(config)#security rsn akm psk enable	#配置WAP2的加密模式为PSK
AP(config)#security rsn akm psk set-key ascii 12345678	#配置PSK共享密码

任务15：IPSec over Gre VPN配置

1. 任务具体描述

为保障学校业务的高可用和高可靠性，需要为业务流量配置备份链路，但又要保障数据

第6章 网络工程项目规划与搭建

的安全性,需要部署站点至站点的VPN技术。两校区之间的专用城域网链路为主链路,实现分校区内网与主校区内网通信。如果主链路出现故障,则需要通过互联网使用IPSec VPN技术进行通信,并且要求GRE隧道学习到分公司内网的路由。网络拓扑结构如图6-1所示。

2. 网络设备配置

(1) RT1路由器配置

```
RT1(config)#interface Tunnel 0                                          #建立隧道接口
RT1(config-if)#ip address 10.10.10.1 255.255.255.0                      #指定隧道接口地址
RT1(config-if)#tunnel source 68.1.1.2                                   #指定隧道源地址
RT1(config-if)#tunnel destination 72.1.1.2                              #指定隧道目的地址
RT1(config-if)#exit
RT1(config)#ip access-list extended 103                                 #定义访问列表
RT1(config-ext-nacl)#permit gre host 68.1.1.2 host 72.1.1.2             #用于加密隧道数据流
RT1(config-ext-nacl)#exit
RT1(config)#crypto isakmp policy 1                                      #配置第一阶段加密策略
RT1(config-isakmp)#encryption 3des                                      #指定加密算法
RT1(config-isakmp)#authentication pre-share                             #使用预共享密钥进行认证
RT1(config-isakmp)#hash sha                                             #配置散列算法
RT1(config-isakmp)#exit
RT1(config)#crypto isakmp key 7 123456 address 72.1.1.2                 #配置加密密钥
RT1(config)#crypto ipsec transform-set gre_tr ah-md5-hmac               #配置第二阶段加密策略
RT1(cfg-crypto-trans)#mode transport                                    #隧道模式为传输模式
RT1(config-crypto-map)#crypto map 1 1 ipsec-isakmp                      #配置多个MAP条目
RT1(config-crypto-map)#set peer 72.1.1.2                                #指定对等体IP
RT1(config-crypto-map)#set transform-set gre_tr                         #引用IPsec的变换集
RT1(config-crypto-map)#match address 103                                #对GRE数据流进行保护
RT1(config-crypto-map)#exit
RT1(config)#router ospf 1                                               #启用OSPF路由进程
RT1(config-router)#network 10.10.10.0 0.0.0.255 area 10                 #宣告路由
RT1(config-router)#exit
RT1(config)#interface Serial 4/0
RT1(config-if-Serial 4/0)#crypto map 1                                  #应用加密映射
```

(2) RT3路由器配置

```
RT3(config)#interface Tunnel 0                                          #建立隧道接口
RT3(config-if)#ip address 10.10.10.2 255.255.255.0                      #指定隧道接口地址
RT3(config-if)#tunnel source 72.1.1.2                                   #指定隧道源地址
RT3(config-if)#tunnel destination 68.1.1.2                              #指定隧道目的地址
RT3(config)#ip access-list extended 103                                 #定义访问列表
RT3(config-ext-nacl)#permit gre host 72.1.1.2 host 68.1.1.2             #用于加密隧道数据流
RT3(config-ext-nacl)#exit
RT3(config)#crypto isakmp policy 1                                      #配置第一阶段加密策略
RT3(config-isakmp)#encryption 3des                                      #指定加密算法
RT3(config-isakmp)#authentication pre-share                             #使用预共享密钥进行认证
RT3(config-isakmp)#hash sha                                             #配置散列算法
RT3(config-isakmp)#exit
RT3(config)#crypto isakmp key 7 123456 address 68.1.1.2                 #配置加密密钥
RT3(config)#crypto ipsec transform-set gre_tr ah-md5-hmac               #配置第二阶段加密策略
RT3(cfg-crypto-trans)#mode transport                                    #隧道模式为传输模式
RT3(cfg-crypto-trans)#exit
```

```
RT3(config)#crypto map 1 1 ipsec-isakmp                    #配置多个MAP条目
RT3(config-isakmp)#set peer 68.1.1.2                       #指定对等体IP
RT3(config-isakmp)#set transform-set gre_tr                #引用IPsec的变换集
RT3(config-isakmp)#match address 103                       #对GRE数据流进行保护
RT3(config-isakmp)#exit
RT3(config)#router ospf 1                                  #启用OSPF路由进程
RT3(config-router)#network 10.10.10.0 0.0.0.255 area 10    #宣告路由
RT3(config-router)#exit
RT3(config)#interface Serial 2/0
RT3(config-if-Serial 2/0)#crypto map 1                     #应用加密映射
```

任务16：PPTP VPN配置

1. 任务具体描述

为了方便学校教师远程办公访问内部网络资源，并保障其安全性，需要部署远程接入VPN技术，在路由器的外部接口配置远程接入VPN，允许该校教师采用PPTP访问内网资源，假设只允许用户teacher1、teacher2、teacher3、teacher4和teacher5访问内部网络，其密码为用户名，其获取的IP地址为192.168.30.150～192.168.30.200。网络拓扑结构如图6-1所示。

2. 网络设备配置

RT1路由器配置

```
RT1(config)#username teacher1 password teacher1            #配置用户信息
RT1(config)#username teacher2 password teacher2            #配置用户信息
RT1(config)#username teacher3 password teacher3            #配置用户信息
RT1(config)#username teacher4 password teacher4            #配置用户信息
RT1(config)#username teacher5 password teacher5            #配置用户信息
RT1(config)#ip local pool vpn 192.168.30.150 192.168.30.200   #配置拨号地址池
RT1(config)#vpdn enable                                    #开启VPN拨入
RT1(config-vpdn)#vpdn-group pptp                           #设置拨号组的名称
RT1(config-vpdn)#!Default PPTP VPDN group
RT1(config-vpdn)#accept-dialin                             #允许远程客户端拨入
RT1(config-vpdn-acc-in)#protocol pptp                      #选用协议进行拨号
RT1(config-vpdn-acc-in)#virtual-template 1                 #设置使用的虚拟模板
RT1(config-vpdn-acc-in)#exit
RT1(config)#interface Virtual-Template 1                   #定义虚拟模板1
RT1(config-if)#ppp authentication chap                     #启用PPP认证
RT1(config-if)#ip unnumbered FastEthernet 0/1              #指定拨入地址
RT1(config-if)#peer default ip address pool vpn            #设置对端IP地址
```

任务17：QoS配置

1. 任务具体描述

为了保障服务器正常工作，防止采用恶意攻击，需要限制服务器群中的服务器向外发送的最大流量不超过2Mbit/s，并且对超出规格的流量做丢弃处理。网络拓扑结构如图6-1所示。

2. 网络设备配置

```
SW2(config)#interface range FastEthernet 0/10-15           #进入服务器所连接的接口
SW2(config-if-range)#rate-limit input 2000000 3000 3000 conform-action continue exceed-action drop
```

第6章　网络工程项目规划与搭建

                           #配置最大流量为512Kbit/s

任务18：远程登录配置

1. 任务具体描述

给所有三层设备配置Telnet/Enable密码，且只允许192.168.100.240～192.168.100.248网段的管理人员远程登录控制，Telnet密码为123456，Enable密码为654321，密码以明文方式存储。网络拓扑结构如图6-1所示。

2. 网络设备配置

（1）SW1交换机配置

1）在SW1交换机上配置远程登录。

```
SW1(config)#enable password 0 654321              #配置enable密码
SW1(config)#line vty 0 35                         #进入线程配置模式
SW1(config-line)#password 123456                  #配置Telnet密码
SW1(config-line)#login                            #设置Telnet登录时身份验证
```

2）在SW1交换机上配置访问控制列表，限制VTY访问。

```
SW1(config)#ip access-list standard 1             #配置访问列表
SW1(config-ext-nacl)#permit host 192.168.100.240  #配置允许访问的用户
SW1(config-ext-nacl)#permit host 192.168.100.241  #配置允许访问的用户
SW1(config-ext-nacl)#permit host 192.168.100.242  #配置允许访问的用户
SW1(config-ext-nacl)#permit host 192.168.100.243  #配置允许访问的用户
SW1(config-ext-nacl)#permit host 192.168.100.244  #配置允许访问的用户
SW1(config-ext-nacl)#permit host 192.168.100.245  #配置允许访问的用户
SW1(config-ext-nacl)#permit host 192.168.100.246  #配置允许访问的用户
SW1(config-ext-nacl)#permit host 192.168.100.247  #配置允许访问的用户
SW1(config-ext-nacl)#permit host 192.168.100.248  #配置允许访问的用户
SW1(config-ext-nacl)#exit
SW1(config)#line vty 0 35                         #进入线程配置模式
SW1(config-line)#access-class 1 in                #进行VTY访问限制
```

（2）SW2交换机配置

1）在SW2交换机上配置远程登录。

```
SW2(config)#enable password 0 654321              #配置enable密码
SW2(config)#line vty 0 35                         #进入线程配置模式
SW2(config-line)#password 123456                  #配置Telnet密码
SW2(config-line)#login                            #设置Telnet登录时身份验证
```

2）在SW2交换机上配置访问控制列表，限制VTY访问。

```
SW2(config)#ip access-list standard 1             #配置访问列表
SW2(config-ext-nacl)#permit host 192.168.100.240  #配置允许访问的用户
SW2(config-ext-nacl)#permit host 192.168.100.241  #配置允许访问的用户
SW2(config-ext-nacl)#permit host 192.168.100.242  #配置允许访问的用户
SW2(config-ext-nacl)#permit host 192.168.100.243  #配置允许访问的用户
SW2(config-ext-nacl)#permit host 192.168.100.244  #配置允许访问的用户
SW2(config-ext-nacl)#permit host 192.168.100.245  #配置允许访问的用户
SW2(config-ext-nacl)#permit host 192.168.100.246  #配置允许访问的用户
SW2(config-ext-nacl)#permit host 192.168.100.247  #配置允许访问的用户
SW2(config-ext-nacl)#permit host 192.168.100.248  #配置允许访问的用户
```

SW2(config-ext-nacl)#exit
SW2(config)#line vty 0 35   #进入线程配置模式
SW2(config-line)#access-class 1 in   #进行VTY访问限制

（3）SW3交换机配置

1）在SW3交换机上配置远程登录。

SW3(config)#enable password 0 654321
SW3(config)#line vty 0 35
SW3(config-line)#password 123456
SW3(config-line)#login

2）在SW3交换机上配置访问控制列表，限制VTY访问。

SW3(config)#ip access-list standard 1
SW3(config-ext-nacl)#permit host 192.168.100.240
SW3(config-ext-nacl)#permit host 192.168.100.241
SW3(config-ext-nacl)#permit host 192.168.100.242
SW3(config-ext-nacl)#permit host 192.168.100.243
SW3(config-ext-nacl)#permit host 192.168.100.244
SW3(config-ext-nacl)#permit host 192.168.100.245
SW3(config-ext-nacl)#permit host 192.168.100.246
SW3(config-ext-nacl)#permit host 192.168.100.247
SW3(config-ext-nacl)#permit host 192.168.100.248
SW3(config)#line vty 0 35
SW3(config-line)#access-class 1 in

（4）SW4交换机配置

1）在SW4交换机上配置远程登录。

SW4(config)#enable password 0 654321
SW4(config)#line vty 0 35
SW4(config-line)#password 123456
SW4(config-line)#login

2）在SW4交换机上配置访问控制列表，限制VTY访问。

SW4(config)#ip access-list standard 1
SW4(config-ext-nacl)#permit host 192.168.100.240
SW4(config-ext-nacl)#permit host 192.168.100.241
SW4(config-ext-nacl)#permit host 192.168.100.242
SW4(config-ext-nacl)#permit host 192.168.100.243
SW4(config-ext-nacl)#permit host 192.168.100.244
SW4(config-ext-nacl)#permit host 192.168.100.245
SW4(config-ext-nacl)#permit host 192.168.100.246
SW4(config-ext-nacl)#permit host 192.168.100.247
SW4(config-ext-nacl)#permit host 192.168.100.248
SW4(config-ext-nacl)#exit
SW4(config)#line vty 0 35
SW4(config-line)#access-class 1 in

（5）RT1路由器配置

1）在RT1路由器上配置远程登录。

RT1(config)#enable password 0 654321
RT1(config)#line vty 0 35
RT1(config-line)#password 123456
RT1(config-line)#login

2）在RT1路由器上配置访问控制列表，限制VTY访问。
RT1(config)#ip access-list standard 1
RT1(config-ext-nacl)#permit host 192.168.100.240
RT1(config-ext-nacl)#permit host 192.168.100.241
RT1(config-ext-nacl)#permit host 192.168.100.242
RT1(config-ext-nacl)#permit host 192.168.100.243
RT1(config-ext-nacl)#permit host 192.168.100.244
RT1(config-ext-nacl)#permit host 192.168.100.245
RT1(config-ext-nacl)#permit host 192.168.100.246
RT1(config-ext-nacl)#permit host 192.168.100.247
RT1(config-ext-nacl)#permit host 192.168.100.248
RT1(config-ext-nacl)#exit
RT1(config)#line vty 0 35
RT1(config-line)#access-class 1 in

（6）RT2路由器配置

1）在RT2路由器上配置远程登录。
RT2(config)#enable password 0 654321
RT2(config)#line vty 0 35
RT2(config-line)#password 123456
RT2(config-line)#login

2）在RT2路由器上配置访问控制列表，限制VTY访问。
RT2(config)#ip access-list standard 1
RT2(config-ext-nacl)#permit host 192.168.100.240
RT2(config-ext-nacl)#permit host 192.168.100.241
RT2(config-ext-nacl)#permit host 192.168.100.242
RT2(config-ext-nacl)#permit host 192.168.100.243
RT2(config-ext-nacl)#permit host 192.168.100.244
RT2(config-ext-nacl)#permit host 192.168.100.245
RT2(config-ext-nacl)#permit host 192.168.100.246
RT2(config-ext-nacl)#permit host 192.168.100.247
RT2(config-ext-nacl)#permit host 192.168.100.248
RT2(config-ext-nacl)#exit
RT2(config)#line vty 0 35
RT2(config-line)#access-class 1 in

（7）RT3路由器配置

1）在RT3路由器上配置远程登录。
RT3(config)#enable password 0 654321
RT3(config)#line vty 0 35
RT3(config-line)#password 123456
RT3(config-line)#login

2）在RT3路由器上配置访问控制列表，限制VTY访问。
RT3(config)#ip access-list standard 1
RT3(config-ext-nacl)#permit host 192.168.100.240
RT3(config-ext-nacl)#permit host 192.168.100.241
RT3(config-ext-nacl)#permit host 192.168.100.242
RT3(config-ext-nacl)#permit host 192.168.100.243
RT3(config-ext-nacl)#permit host 192.168.100.244
RT3(config-ext-nacl)#permit host 192.168.100.245

RT3(config-ext-nacl)#permit host 192.168.100.246
RT3(config-ext-nacl)#permit host 192.168.100.247
RT3(config-ext-nacl)#permit host 192.168.100.248
RT3(config-ext-nacl)#exit
RT3(config)#line vty 0 35
RT3(config-line)#access-class 1 in

参考配置

1. RT1路由器参考配置

RT1#show run

Building configuration...
Current configuration : 4768 bytes
!
version RGOS 10.3(5b1), Release(84749)(Thu May 13 09:09:02 CST 2010 -ngcf66)
hostname RT1
!
!
route-map jiaoyu permit 10
 match ip address 3
 set ip next-hop 210.21.1.1
!
username teacher1 password teacher1
username teacher2 password teacher2
username teacher3 password teacher3
username teacher4 password teacher4
username teacher5 password teacher5
no service password-encryption
!
!
ip access-list standard 1
 10 permit host 192.168.100.240
 20 permit host 192.168.100.241
 30 permit host 192.168.100.242
 40 permit host 192.168.100.243
 50 permit host 192.168.100.244
 60 permit host 192.168.100.245
 70 permit host 192.168.100.246
 80 permit host 192.168.100.247
 90 permit host 192.168.100.248
!
!
ip access-list standard 3
 10 permit 192.168.10.0 0.0.0.255
 20 permit 192.168.20.0 0.0.0.255
!
!

```
ip access-list extended 100
  10 permit ip any any
!
!
ip access-list extended 101
  10 permit ip any any
!
!
ip access-list extended 102
  10 permit tcp any host 68.1.1.10
!
!
ip access-list extended 103
  10 permit gre host 68.1.1.2 host 72.1.1.2
!
!
ip local pool vpn 192.168.30.150 192.168.30.200
!
!
enable password 654321
!
crypto isakmp policy 1
  encryption 3des
  authentication pre-share
  hash sha
!
!
crypto isakmp key 7 1443185e75477d address 72.1.1.2
crypto ipsec transform-set gre_tr ah-md5-hmac
  mode transport
crypto map 1 1 ipsec-isakmp
  set peer 72.1.1.2
  set transform-set gre_tr
  match address 103
!
!
vpdn enable
!
vpdn-group pptp
! Default PPTP VPDN group
  accept-dialin
    protocol pptp
    virtual-template 1
!
!
interface Serial 2/0
  ip nat inside
  ip ospf authentication message-digest
  ip ospf message-digest-key 1 md5 zhuxiaoqu
  ip address 10.1.1.9 255.255.255.252
```

```
  clock rate 64000
!
interface Serial 3/0
  ip nat outside
  ip address 210.21.1.2 255.255.255.0
  clock rate 64000
!
interface Serial 4/0
  ip nat outside
  ip address 68.1.1.2 255.255.255.240
  crypto map 1
  clock rate 64000
!
interface FastEthernet 0/0
  ip nat inside
  ip policy route-map jiaoyu
  ip ospf authentication message-digest
  ip ospf message-digest-key 1 md5 zhuxiaoqu
  ip address 10.1.1.1 255.255.255.252
  duplex auto
  speed auto
!
interface FastEthernet 0/1
  ip nat inside
  ip ospf authentication message-digest
  ip ospf message-digest-key 1 md5 zhuxiaoqu
  ip address 10.1.1.5 255.255.255.252
  duplex auto
  speed auto
!
interface FastEthernet 0/2
  duplex auto
  speed auto
!
interface Virtual-Template 1
  ppp authentication chap
  ip unnumbered FastEthernet 0/1
  peer default ip address pool vpn
!
interface Tunnel 0
  ip address 10.10.10.1 255.255.255.0
  tunnel source 68.1.1.2
  tunnel destination 72.1.1.2
!
ip nat pool dianxin 68.1.1.3 68.1.1.5 netmask 255.255.255.248
ip nat pool jiaoyu 210.21.1.10 210.21.1.15 netmask 255.255.255.0
ip nat pool web 192.168.100.5 192.168.100.6 prefix-length 24 type rotary
ip nat inside source static tcp 192.168.100.4 20 68.1.1.8 20
ip nat inside source static tcp 192.168.100.4 21 68.1.1.8 21
ip nat inside source list 100 pool dianxin overload
```

第6章 网络工程项目规划与搭建

```
ip nat inside source list 101 pool jiaoyu overload
ip nat inside destination list 102 pool web
!
!
router ospf 1
  router-id 3.3.3.3
  network 10.1.1.0 0.0.0.3 area 0
  network 10.1.1.4 0.0.0.3 area 0
  network 10.1.1.8 0.0.0.3 area 0
  network 10.10.10.0 0.0.0.255 area 10
  default-information originate
!
!
ip route 0.0.0.0 0.0.0.0 68.1.1.1
ip route 211.1.0.0 255.255.0.0 210.21.1.1
ip route 212.1.0.0 255.255.255.0 210.21.1.1
ip route 212.2.0.0 255.255.255.0 210.21.1.1
ip route 212.3.0.0 255.255.255.0 210.21.1.1
ip route 212.4.0.0 255.255.255.0 210.21.1.1
ip route 212.5.0.0 255.255.255.0 210.21.1.1
ip route 212.6.0.0 255.255.255.0 210.21.1.1
ip route 212.7.0.0 255.255.255.0 210.21.1.1
ip route 213.1.0.0 255.255.255.0 210.21.1.1
ip route 213.1.1.0 255.255.255.0 210.21.1.1
ip route 213.1.2.0 255.255.255.0 210.21.1.1
ip route 213.1.3.0 255.255.255.0 210.21.1.1
ip route 213.1.4.0 255.255.255.0 210.21.1.1
ip route 213.1.5.0 255.255.255.0 210.21.1.1
ip route 213.1.6.0 255.255.255.0 210.21.1.1
ip route 213.1.7.0 255.255.255.0 210.21.1.1
ip route 213.1.8.0 255.255.255.0 210.21.1.1
ip route 213.1.9.0 255.255.255.0 210.21.1.1
ip route 213.1.10.0 255.255.255.0 210.21.1.1
ip route 213.1.11.0 255.255.255.0 210.21.1.1
ip route 213.1.12.0 255.255.255.0 210.21.1.1
ip route 213.1.13.0 255.255.255.0 210.21.1.1
ip route 213.1.14.0 255.255.255.0 210.21.1.1
ip route 213.1.15.0 255.255.255.0 210.21.1.1
!
!
ref parameter 50 400
line con 0
line aux 0
line vty 0 35
  access-class 1 in
  login
  password 123456
!
```

!
end
RT1(config)#

2. RT2路由器参考配置

RT2#show run

Building configuration...
Current configuration : 1730 bytes
!
version RGOS 10.3(5b1), Release(84749)(Thu May 13 09:09:02 CST 2010 -ngcf66)
hostname RT2
!
!
username fenxiaoqu password 123456
no service password-encryption
!
!
ip access-list standard 1
 10 permit host 192.168.100.240
 20 permit host 192.168.100.241
 30 permit host 192.168.100.242
 40 permit host 192.168.100.243
 50 permit host 192.168.100.244
 60 permit host 192.168.100.245
 70 permit host 192.168.100.246
 80 permit host 192.168.100.247
 90 permit host 192.168.100.248
!
!
enable password 654321
!
!
interface Serial 2/0
 ip ospf authentication message-digest
 ip ospf message-digest-key 1 md5 zhuxiaoqu
 ip address 10.1.1.10 255.255.255.252
 clock rate 64000
!
interface Serial 3/0
 encapsulation PPP
 ppp authentication pap chap
 ip ospf authentication message-digest
 ip ospf message-digest-key 1 md5 zhuxiaoqu
 ip address 10.1.1.25 255.255.255.252
!
interface FastEthernet 0/0
 ip ospf authentication message-digest
 ip ospf message-digest-key 1 md5 zhuxiaoqu

第6章 网络工程项目规划与搭建

```
  ip address 10.1.1.13 255.255.255.252
  duplex auto
  speed auto
!
interface FastEthernet 0/1
  ip ospf authentication message-digest
  ip ospf message-digest-key 1 md5 zhuxiaoqu
  ip address 10.1.1.17 255.255.255.252
  duplex auto
  speed auto
!
interface FastEthernet 0/2
  duplex auto
  speed auto
!
!
router ospf 1
  router-id 4.4.4.4
  network 10.1.1.8 0.0.0.3 area 0
  network 10.1.1.12 0.0.0.3 area 0
  network 10.1.1.16 0.0.0.3 area 0
  network 10.1.1.24 0.0.0.3 area 0
!
!
ref parameter 50 400
line con 0
line aux 0
line vty 0 35
  access-class 1 in
  login
  password 123456
!
!
end
RT2(config)#
```

3. RT3路由器参考配置

```
RT3#show run

Building configuration...
Current configuration : 2679 bytes

!
version RGOS 10.3(5b1), Release(84749)(Thu May 13 09:09:02 CST 2010 -ngcf66)
hostname RT3
!
!
time-range 1
  periodic Weekdays 8:00 to 17:00
!
```

key chain rip
 key 1
   key-string fenxiaoqu
!
no service password-encryption
!
!
ip access-list standard 1
 10 permit host 192.168.100.240
 20 permit host 192.168.100.241
 30 permit host 192.168.100.242
 40 permit host 192.168.100.243
 50 permit host 192.168.100.244
 60 permit host 192.168.100.245
 70 permit host 192.168.100.246
 80 permit host 192.168.100.247
 90 permit host 192.168.100.248
!
!
ip access-list extended 101
 10 permit ip 192.168.200.0 0.0.0.255 any time-range 1
 20 deny ip 192.168.200.0 0.0.0.255 any
 30 permit ip any any
!
!
ip access-list extended 103
 10 permit gre host 72.1.1.2 host 68.1.1.2
!
!
enable password 654321
!
crypto isakmp policy 1
  encryption 3des
  authentication pre-share
  hash sha
!
!
crypto isakmp key 7 08527d46467253 address 68.1.1.2
crypto ipsec transform-set gre_tr ah-md5-hmac
  mode transport
crypto map 1 1 ipsec-isakmp
  set peer 68.1.1.2
  set transform-set gre_tr
  match address 103
!
!
interface Serial 2/0
  ip nat outside
  ip address 72.1.1.2 255.255.255.252
  crypto map 1

```
    clock rate 64000
!
interface Serial 3/0
  encapsulation PPP
  ppp chap hostname fenxiaoqu
  ppp chap password 123456
  ppp pap sent-username fenxiaoqu password 123456
  ip nat inside
  ip ospf authentication message-digest
  ip ospf message-digest-key 1 md5 zhuxiaoqu
  ip address 10.1.1.26 255.255.255.252
!
interface FastEthernet 0/0
  ip nat inside
  ip rip authentication mode md5
  ip rip authentication key-chain rip
  ip address 10.1.1.29 255.255.255.252
  duplex auto
  speed auto
!
interface FastEthernet 0/1
  ip nat inside
  ip rip authentication mode md5
  ip rip authentication key-chain rip
  ip address 10.1.1.33 255.255.255.252
  duplex auto
  speed auto
!
interface FastEthernet 0/2
  duplex auto
  speed auto
!
interface Tunnel 0
  ip address 10.10.10.2 255.255.255.0
  tunnel source 72.1.1.2
  tunnel destination 68.1.1.2
!
ip nat inside source list 101 interface Serial 2/0
!
!
router ospf 1
  router-id 5.5.5.5
  redistribute rip subnets
  network 10.1.1.24 0.0.0.3 area 10
  network 10.10.10.0 0.0.0.255 area 10
  default-information originate
!
!
router rip
  version 2
```

```
 network 10.0.0.0
 no auto-summary
 redistribute ospf 1 metric 2
!
!
ip route 0.0.0.0 0.0.0.0 72.1.1.1
!
!
ref parameter 50 400
line con 0
line aux 0
line vty 0 35
 access-class 1 in
 login
 password 123456
!
!
end
```

4. SW1交换机参考配置

```
SW1#show run

Building configuration...
Current configuration : 3986 bytes
!
version RGOS 10.4(2) Release(75955)(Mon Jan 25 19:33:15 CST 2010 -ngcf31)
hostname SW1
!
!
nfpp
!
!
vlan 1
!
vlan 10
 name jiaoxuezhongxin
!
vlan 20
 name shixunzhongxin
!
!
no service password-encryption
!
!
ip access-list standard 1
 10 permit host 192.168.100.240
 20 permit host 192.168.100.241
 30 permit host 192.168.100.242
 40 permit host 192.168.100.243
 50 permit host 192.168.100.244
```

```
   60 permit host 192.168.100.245
   70 permit host 192.168.100.246
   80 permit host 192.168.100.247
   90 permit host 192.168.100.248
!
!
enable password 654321
!
!
interface FastEthernet 0/1
 no switchport
 ip ospf authentication message-digest
 ip ospf message-digest-key 1 md5 zhuxiaoqu
 no ip proxy-arp
 ip address 10.1.1.2 255.255.255.252
!
interface FastEthernet 0/2
 no switchport
 ip ospf authentication message-digest
 ip ospf message-digest-key 1 md5 zhuxiaoqu
 no ip proxy-arp
 ip address 10.1.1.14 255.255.255.252
!
interface FastEthernet 0/3
!
interface FastEthernet 0/4
!
interface FastEthernet 0/5
!
interface FastEthernet 0/6
 switchport access vlan 10
 switchport port-security maximum 1
 switchport port-security violation shutdown
 switchport port-security
!
interface FastEthernet 0/7
 switchport access vlan 10
 switchport port-security maximum 1
 switchport port-security violation shutdown
 switchport port-security
!
interface FastEthernet 0/8
 switchport access vlan 10
 switchport port-security maximum 1
 switchport port-security violation shutdown
 switchport port-security
!
interface FastEthernet 0/9
 switchport access vlan 10
 switchport port-security maximum 1
```

```
  switchport port-security violation shutdown
  switchport port-security
!
interface FastEthernet 0/10
  switchport access vlan 10
  switchport port-security maximum 1
  switchport port-security violation shutdown
  switchport port-security
!
interface FastEthernet 0/11
  switchport access vlan 20
  switchport port-security maximum 1
  switchport port-security violation shutdown
  switchport port-security
!
interface FastEthernet 0/12
  switchport access vlan 20
  switchport port-security maximum 1
  switchport port-security violation shutdown
  switchport port-security
!
interface FastEthernet 0/13
  switchport access vlan 20
  switchport port-security maximum 1
  switchport port-security violation shutdown
  switchport port-security
!
interface FastEthernet 0/14
  switchport access vlan 20
  switchport port-security maximum 1
  switchport port-security violation shutdown
  switchport port-security
!
interface FastEthernet 0/15
  switchport access vlan 20
  switchport port-security maximum 1
  switchport port-security violation shutdown
  switchport port-security
!
interface FastEthernet 0/16
!
interface FastEthernet 0/17
!
interface FastEthernet 0/18
!
interface FastEthernet 0/19
!
interface FastEthernet 0/20
!
interface FastEthernet 0/21
```

第6章 网络工程项目规划与搭建

```
!
interface FastEthernet 0/22
!
interface FastEthernet 0/23
 no switchport
 port-group 1
!
interface FastEthernet 0/24
 no switchport
 port-group 1
!
interface GigabitEthernet 0/25
!
interface GigabitEthernet 0/26
!
interface AggregatePort 1
 no switchport
 ip ospf authentication message-digest
 ip ospf message-digest-key 1 md5 zhuxiaoqu
 no ip proxy-arp
 ip address 10.1.1.21 255.255.255.252
!
interface VLAN 10
 no ip proxy-arp
 ip address 192.168.10.1 255.255.255.0
!
interface VLAN 20
 no ip proxy-arp
 ip address 192.168.20.1 255.255.255.0
!
!
aggregateport load-balance src-ip
!
!
router ospf 1
 router-id 1.1.1.1
 network 10.1.1.0 0.0.0.3 area 0
 network 10.1.1.12 0.0.0.3 area 0
 network 10.1.1.20 0.0.0.3 area 0
 network 192.168.10.0 0.0.0.255 area 0
 network 192.168.20.0 0.0.0.255 area 0
!
!
line con 0
line vty 0 35
 access-class 1 in
 login
 password 123456
!
!
end
```

5. SW2交换机参考配置

SW2#show run

Building configuration...
Current configuration : 4692 bytes
!
version RGOS 10.4(2) Release(75955)(Mon Jan 25 19:33:15 CST 2010 -ngcf31)
hostname SW2
!
!
time-range 1
 periodic Weekdays 8:00 to 17:00
!
!
nfpp
!
!
vlan 1
!
vlan 30
 name xingzhengbangongzhongxin
!
vlan 100
 name xinxizhongxin
!
!
no service password-encryption
service dhcp
!
ip dhcp excluded-address 192.168.30.1
!
ip dhcp pool vlan30
 network 192.168.30.0 255.255.255.0
 dns-server 210.21.1.66
 default-router 192.168.30.1
!
!
ip access-list standard 1
 10 permit host 192.168.100.240
 20 permit host 192.168.100.241
 30 permit host 192.168.100.242
 40 permit host 192.168.100.243
 50 permit host 192.168.100.244
 60 permit host 192.168.100.245
 70 permit host 192.168.100.246
 80 permit host 192.168.100.247
 90 permit host 192.168.100.248
!
!
ip access-list extended 100

10 permit tcp any host 192.168.100.50 eq 2020 time-range 1
   20 permit udp any host 192.168.100.50 eq 2020 time-range 1
   30 deny udp any host 192.168.100.50 eq 2020
   40 deny tcp any host 192.168.100.50 eq 2020
   50 permit ip any any
!
!
enable password 654321
!
!
interface FastEthernet 0/1
  no switchport
  ip ospf authentication message-digest
  ip ospf message-digest-key 1 md5 zhuxiaoqu
  no ip proxy-arp
  ip address 10.1.1.6 255.255.255.252
!
interface FastEthernet 0/2
  no switchport
  ip ospf authentication message-digest
  ip ospf message-digest-key 1 md5 zhuxiaoqu
  no ip proxy-arp
  ip address 10.1.1.18 255.255.255.252
!
interface FastEthernet 0/3
!
interface FastEthernet 0/4
!
interface FastEthernet 0/5
!
interface FastEthernet 0/6
  switchport access vlan 30
  switchport port-security maximum 1
  switchport port-security violation shutdown
  switchport port-security
!
interface FastEthernet 0/7
  switchport access vlan 30
  switchport port-security maximum 1
  switchport port-security violation shutdown
  switchport port-security
!
interface FastEthernet 0/8
  switchport access vlan 30
  switchport port-security maximum 1
  switchport port-security violation shutdown
  switchport port-security
!
interface FastEthernet 0/9
  switchport access vlan 30

switchport port-security maximum 1
　　switchport port-security violation shutdown
　　switchport port-security
!
interface FastEthernet 0/10
　　switchport access vlan 30
　　switchport port-security maximum 1
　　switchport port-security violation shutdown
　　switchport port-security
　　rate-limit input 512 256
!
interface FastEthernet 0/11
　　switchport access vlan 100
　　switchport port-security maximum 1
　　switchport port-security violation shutdown
　　switchport port-security
　　rate-limit input 512 256
!
interface FastEthernet 0/12
　　switchport access vlan 100
　　switchport port-security maximum 1
　　switchport port-security violation shutdown
　　switchport port-security
　　rate-limit input 512 256
!
interface FastEthernet 0/13
　　switchport access vlan 100
　　switchport port-security maximum 1
　　switchport port-security violation shutdown
　　switchport port-security
　　rate-limit input 512 256
!
interface FastEthernet 0/14
　　switchport access vlan 100
　　switchport port-security maximum 1
　　switchport port-security violation shutdown
　　switchport port-security
　　rate-limit input 512 256
!
interface FastEthernet 0/15
　　switchport access vlan 100
　　switchport port-security maximum 1
　　switchport port-security violation shutdown
　　switchport port-security
　　rate-limit input 512 256
!
interface FastEthernet 0/16
!
interface FastEthernet 0/17
!

第6章 网络工程项目规划与搭建

```
interface FastEthernet 0/18
!
interface FastEthernet 0/19
!
interface FastEthernet 0/20
!
interface FastEthernet 0/21
!
interface FastEthernet 0/22
!
interface FastEthernet 0/23
 no switchport
 port-group 1
!
interface FastEthernet 0/24
 no switchport
 port-group 1
!
interface GigabitEthernet 0/25
!
interface GigabitEthernet 0/26
!
interface AggregatePort 1
 no switchport
 ip ospf authentication message-digest
 ip ospf message-digest-key 1 md5 zhuxiaoqu
 no ip proxy-arp
 ip address 10.1.1.22 255.255.255.252
!
interface VLAN 30
 no ip proxy-arp
 ip address 192.168.30.1 255.255.255.0
!
interface VLAN 100
 no ip proxy-arp
 ip access-group 100 out
 ip address 192.168.100.1 255.255.255.0
!
!
aggregateport load-balance src-ip
!
!
router ospf 1
 router-id 2.2.2.2
 network 10.1.1.4 0.0.0.3 area 0
 network 10.1.1.16 0.0.0.3 area 0
 network 10.1.1.20 0.0.0.3 area 0
 network 192.168.30.0 0.0.0.255 area 0
 network 192.168.100.0 0.0.0.255 area 0
!
```

!
line con 0
line vty 0 35
　access-class 1 in
　login
　password 123456
!
!
end

6. SW3交换机参考配置

SW3#show run

Building configuration...
Current configuration : 3677 bytes

!
version RGOS 10.4(2) Release(75955)(Mon Jan 25 19:33:15 CST 2010 -ngcf31)
hostname SW3
!
!
nfpp
!
!
vlan 1
!
vlan 40
　name jiaoxuezhongxin
!
vlan 50
　name shixunzhongxin
!
!
key chain rip
　key 1
　　key-string fenxiaoqu
!
no service password-encryption
!
!
ip access-list standard 1
　10 permit host 192.168.100.240
　20 permit host 192.168.100.241
　30 permit host 192.168.100.242
　40 permit host 192.168.100.243
　50 permit host 192.168.100.244
　60 permit host 192.168.100.245
　70 permit host 192.168.100.246
　80 permit host 192.168.100.247
　90 permit host 192.168.100.248

第6章　网络工程项目规划与搭建

```
!
!
enable password 654321
!
!
interface FastEthernet 0/1
 no switchport
 ip rip authentication mode md5
 ip rip authentication key-chain rip
 no ip proxy-arp
 ip address 10.1.1.30 255.255.255.252
!
interface FastEthernet 0/2
!
interface FastEthernet 0/3
!
interface FastEthernet 0/4
!
interface FastEthernet 0/5
!
interface FastEthernet 0/6
 switchport access vlan 40
 switchport port-security maximum 1
 switchport port-security violation shutdown
 switchport port-security
!
interface FastEthernet 0/7
 switchport access vlan 40
 switchport port-security maximum 1
 switchport port-security violation shutdown
 switchport port-security
!
interface FastEthernet 0/8
 switchport access vlan 40
 switchport port-security maximum 1
 switchport port-security violation shutdown
 switchport port-security
!
interface FastEthernet 0/9
 switchport access vlan 40
 switchport port-security maximum 1
 switchport port-security violation shutdown
 switchport port-security
!
interface FastEthernet 0/10
 switchport access vlan 40
 switchport port-security maximum 1
 switchport port-security violation shutdown
 switchport port-security
!
```

```
interface FastEthernet 0/11
  switchport access vlan 50
  switchport port-security maximum 1
  switchport port-security violation shutdown
  switchport port-security
!
interface FastEthernet 0/12
  switchport access vlan 50
  switchport port-security maximum 1
  switchport port-security violation shutdown
  switchport port-security
!
interface FastEthernet 0/13
  switchport access vlan 50
  switchport port-security maximum 1
  switchport port-security violation shutdown
  switchport port-security
!
interface FastEthernet 0/14
  switchport access vlan 50
  switchport port-security maximum 1
  switchport port-security violation shutdown
  switchport port-security
!
interface FastEthernet 0/15
  switchport access vlan 50
  switchport port-security maximum 1
  switchport port-security violation shutdown
  switchport port-security
!
interface FastEthernet 0/16
!
interface FastEthernet 0/17
!
interface FastEthernet 0/18
!
interface FastEthernet 0/19
!
interface FastEthernet 0/20
!
interface FastEthernet 0/21
!
interface FastEthernet 0/22
!
interface FastEthernet 0/23
  no switchport
  port-group 2
!
interface FastEthernet 0/24
  no switchport
```

```
  port-group 2
!
interface GigabitEthernet 0/25
!
interface GigabitEthernet 0/26
!
interface AggregatePort 2
 no switchport
 ip rip authentication mode md5
 ip rip authentication key-chain rip
 no ip proxy-arp
 ip address 10.1.1.41 255.255.255.252
!
interface VLAN 40
 no ip proxy-arp
 ip address 192.168.40.1 255.255.255.0
!
interface VLAN 50
 no ip proxy-arp
 ip address 192.168.50.1 255.255.255.0
!
!
aggregateport load-balance dst-ip
!
!
router rip
 version 2
 network 10.0.0.0
 network 192.168.40.0
 network 192.168.50.0
 no auto-summary
!
!
line con 0
line vty 0 35
 access-class 1 in
 login
 password 123456
!
!
end
```

7. SW4交换机参考配置

```
SW4#show run

Building configuration...
Current configuration : 3708 bytes
!
version RGOS 10.4(2) Release(75955)(Mon Jan 25 19:33:15 CST 2010 -ngcf31)
hostname SW4
```

```
!
nfpp
!
vlan 1
!
vlan 60
  name xingzhengbangongzhongxin
!
key chain rip
  key 1
    key-string fenxiaoqu
!
no service password-encryption
!
!
ip access-list standard 1
  10 permit host 192.168.100.240
  20 permit host 192.168.100.241
  30 permit host 192.168.100.242
  40 permit host 192.168.100.243
  50 permit host 192.168.100.244
  60 permit host 192.168.100.245
  70 permit host 192.168.100.246
  80 permit host 192.168.100.247
  90 permit host 192.168.100.248
!
!
enable password 654321
!
!
interface FastEthernet 0/1
  no switchport
  ip rip authentication mode md5
  ip rip authentication key-chain rip
  no ip proxy-arp
  ip address 10.1.1.34 255.255.255.252
!
interface FastEthernet 0/2
!
interface FastEthernet 0/3
!
interface FastEthernet 0/4
!
interface FastEthernet 0/5
!
interface FastEthernet 0/6
  switchport access vlan 60
  switchport port-security maximum 1
  switchport port-security violation shutdown
  switchport port-security
```

```
!
interface FastEthernet 0/7
 switchport access vlan 60
 switchport port-security maximum 1
 switchport port-security violation shutdown
 switchport port-security
!
interface FastEthernet 0/8
 switchport access vlan 60
 switchport port-security maximum 1
 switchport port-security violation shutdown
 switchport port-security
!
interface FastEthernet 0/9
 switchport access vlan 60
 switchport port-security maximum 1
 switchport port-security violation shutdown
 switchport port-security
!
interface FastEthernet 0/10
 switchport access vlan 60
 switchport port-security maximum 1
 switchport port-security violation shutdown
 switchport port-security
!
interface FastEthernet 0/11
 switchport access vlan 60
 switchport port-security maximum 1
 switchport port-security violation shutdown
 switchport port-security
!
interface FastEthernet 0/12
 switchport access vlan 60
 switchport port-security maximum 1
 switchport port-security violation shutdown
 switchport port-security
!
interface FastEthernet 0/13
 switchport access vlan 60
 switchport port-security maximum 1
 switchport port-security violation shutdown
 switchport port-security
!
interface FastEthernet 0/14
 switchport access vlan 60
 switchport port-security maximum 1
 switchport port-security violation shutdown
 switchport port-security
!
interface FastEthernet 0/15
```

switchport access vlan 60
   switchport port-security maximum 1
   switchport port-security violation shutdown
   switchport port-security
!
interface FastEthernet 0/16
!
interface FastEthernet 0/17
!
interface FastEthernet 0/18
!
interface FastEthernet 0/19
!
interface FastEthernet 0/20
!
interface FastEthernet 0/21
!
interface FastEthernet 0/22
  no switchport
  no ip proxy-arp
  ip address 10.1.1.37 255.255.255.252
!
interface FastEthernet 0/23
  no switchport
  port-group 2
!
interface FastEthernet 0/24
  no switchport
  port-group 2
!
interface GigabitEthernet 0/25
!
interface GigabitEthernet 0/26
!
interface AggregatePort 2
  no switchport
  ip rip authentication mode md5
  ip rip authentication key-chain rip
  no ip proxy-arp
  ip address 10.1.1.42 255.255.255.252
!
interface VLAN 60
  no ip proxy-arp
  ip address 192.168.60.1 255.255.255.0
!
!
aggregateport load-balance dst-ip
!
!
router rip

```
 version 2
 network 10.0.0.0
 network 192.168.60.0
no auto-summary
!
!
line con 0
line vty 0 35
 access-class 1 in
 login
 password 123456
!
!
end
```

参 考 文 献

[1] 史蒂文斯. TCP/IP详解[M]. 范建华,等译. 北京:机械工业出版社,2007.
[2] 查普尔,蒂特尔. TCP/IP协议原理与应用[M]. 张长富,等译. 北京:清华大学出版社,2009.
[3] 福罗赞. TCP/IP协议族[M]. 王海,等译. 北京:清华大学出版社,2001.
[4] 塔嫩鲍姆,等. 计算机网络:英文版[M]. 第5版. 北京:机械工业出版社,2011.
[5] 谢希仁. 计算机网络[M]. 北京:电子工业出版社,2008.
[6] 张选波. 企业网络构建与安全管理项目教程[M]. 北京:机械工业出版社,2012.
[7] 爱德华兹,等. CCNP四合一学习指南[M]. 张波,谢琳,译. 北京:电子工业出版社,2005.